Alexander Schmitt · Wolfgang Minker

Towards Adaptive Spoken Dialog Systems

Alexander Schmitt
Institute of Communications Engineering
University of Ulm
Ulm
Germany

Wolfgang Minker
Institute of Communications Engineering
University of Ulm
Ulm
Germany

ISBN 978-1-4614-4592-0 ISBN 978-1-4614-4593-7 (eBook)
DOI 10.1007/978-1-4614-4593-7
Springer New York Heidelberg Dordrecht London

Library of Congress Control Number: 2012944961

Printed on acid-free paper

Springer is part of Springer Science+Business Media (www.springer.com)

Towards Adaptive Spoken Dialog Systems

Preface

This book investigates stochastic methods for the automatic detection of critical dialog situations in Spoken Dialog Systems (SDS) and implements data-driven modeling and prediction techniques. The achievement of this approach is to allow for a robust and user-friendly interaction with next generation SDS.

The advances in spoken language technology have led to an increasing deployment of SDS in the field, such as speech-enabled personal assistants in our smartphones. Limitations of the current technology and the great complexity of natural language interaction between man and machine keep producing problems in communication. We users may particularly experience this in our everyday usage of telephone-based SDS in call centers. Low speech recognition performance, false expectations towards the SDS, and sometimes bad dialog design lead to frustration and dialog crashes.

In this book, we present a data-driven online monitoring approach that enables future SDS to automatically recognize negative dialog patterns. Thereby we implement novel statistical and machine learning-based approaches. Using the knowledge about existing problems in the interaction future dialog systems will be able to change dialog flows dynamically and ultimately solve those problems. Unlike rule-based approaches, the presented statistical procedures allow a more flexible, more portable, and more accurate use.

After an introduction into spoken dialog technology the book describes the foundations of machine learning and pattern recognition, serving as basis for the presented approaches. Related work in the field of emotion recognition and data-driven evaluation of SDS (e.g. the PARADISE approach) are presented and discussed.

The main part of the book begins with a major symptom, which is closely connected to poor communication and critical dialog situations, namely the detection of negative emotions. Related work in the field is frequently based on artificial datasets using acted and enacted speech, which may not be transferred to real-life applications. We investigate how frequently users show negative emotions in real-life SDS and develop novel approaches to speech-based emotion recognition. We follow a multilayer approach to model and detect emotions by using classifiers based on acoustic, linguistic and contextual features. We prove that acoustic models outperform linguistic models in recognizing emotions in real-life SDS. Furthermore, we examine the relationship between the interaction flow and the occurrence of emotions. We may show that the interaction flow of a human–machine dialog has an influence on the user's emotional state. This fact allows us to support emotion recognition using parameters that describe the previous interaction. For our studies, we exclusively employ non-acted recordings from real users.

Not all users react emotionally when problems arise in the interaction with an SDS. In the second step, we therefore present novel statistical methods that allow spotting problems within a dialog-based VI on interaction patterns. The presented Interaction Quality paradigm demonstrates how a continuous quality assessment of ongoing spoken human–machine interaction may be achieved. This expert-based approach represents an objective quality view on an interaction. To which degree this paradigm mirrors subjective user satisfaction is assessed in a laboratory study with 46 users. To our knowledge, this study is the first to assess user satisfaction during SDS interactions. It can be shown that subjective satisfaction correlates with objective quality assessments. Interaction parameters influencing user satisfaction are statistically determined and discussed. Furthermore, we present approaches that will enable future dialog systems to predict the dialog outcome during an interaction. The latter models allow dialog systems in call center applications to escalate endangered calls promptly to call center agents who may help out. Problems specific to the estimation of the dialog outcome are assessed and solved. An open-source workbench supporting the development and evaluation of such statistical models and the presentation of a parameter set used to quantify SDS interactions round off the book. The proposed approaches will increase user-friendliness and robustness in future SDS. All models have been evaluated on several large datasets of commercial and non-commercial SDS and thus have been tested for practical use.

Acknowledgments

This project would not have been possible without the support of many people.

The authors express their deepest gratitude to David Sündermann, Jackson Liscombe, and Roberto Pieraccini from SpeechCycle Inc. (USA) for supporting us all the time with corpora still "hot from recording", their helpful advice, and for their warm welcomes during our stays in New York.

We are notably grateful to Tim Polzehl (Deutsche Telekom Laboratories, Berlin) and Florian Metze (Carnegie Mellon University, Pittsburgh) for our prosperous collaboration in the field of emotion recognition during the past years. We would further like to thank the crew from Deutsche Telekom Laboratories, in particular Sebastian Möller, Klaus-Peter Engelbrecht, and Christine Kühnel for their friendly and collegial cooperation and the exchange of valuable ideas at numerous speech conferences.

We are indebted to our research colleagues, students, and the technical staff at the Dialog Systems group at University of Ulm. In particular we owe a debt of gratitude to Tobias Heinroth, Stefan Ultes, Benjamin Schatz, Shu Ding, Uli Tschaffon, and Carolin Hank as well as Nada Sharaf and Sherief Mowafey from the German University in Cairo (Egypt) for their fruitful discussions and the tireless exchange of new ideas.

Thanks to Allison Michael from Springer for his assistance during the publishing process.

Finally, we thank our families for their encouragement.

Contents

Acronyms

AI	Artificial Intelligence
ANN	Artificial Neural Network
ASR	Automatic Speech Recognition
AVM	Attribute Value Matrix
BOW	Bag-of-Words
CART	Classification and Regression Trees
CFG	Context-free Grammar
DCT	Discrete Cosine Transform
DES	Danish Emotional Speech
DF	Document Frequency
DM	Dialog Manager
DTMF	Dual Tone Multiple Frequency
FN	False Negative
FP	False Positive
GUI	Graphical User Interface
HCI	Human–Computer Interaction
HNR	Harmonics to Noise Ratio
HMIHY	How May I Help You?
HMM	Hidden Markov Model
IDF	Inverse Document Frequency
IGR	Information Gain Ratio
IP	Interaction Parameter
IVR	Interactive Voice Response
IQ	Interaction Quality
IVR	Interactive Voice Response
JDBC	Java Database Connectivity
KNN	*k*-nearest Neighbor
LDC	Linear Discriminative Classification
LOO	Leave-One-Out
LOSO	Leave-One-Speaker-Out
LU	Language Understanding

MAE	Mean Absolute Error
MFCC	Mel-Frequency Cepstral Coefficients
ML	Machine Learning
MR/R	Match-Rate-per-Rate
MSE	Mean Squared Error
PDA	Personal Digital Assistant
POMDP	Partially Observable Markov Decision Process
QoE	Quality of Experience
QoS	Quality of Service
RCP	Rich Client Platform
SC-BAC	SpeechCycle Broadband Agent Corpus
SC-VAC	SpeechCycle Video Agent Corpus
SDS	Spoken Dialog System
SMO	Sequential Minimal Optimization
SQL	Structured Query Language
SVM	Support Vector Machine
TF	Term Frequency
TF.IDF	Term Frequency Inverse Document Frequency
TP	True Positive
TN	True Negative
TTS	Text-To-Speech
UAR	Unweighted Average Recall
US	User Satisfaction
UTD	User Turn Duration
VUI	Voice User Interface
VOD	Video on Demand
WAR	Weighted Average Recall
WER	Word Error Rate

Chapter 1
Introduction

Developing machines that listen and speak has intrigued scientists and engineers for decades. The first speech recognition system was constructed at Bell Laboratories in 1952 by three scientists and was able to recognize sequences of isolated digits of a single speaker (Davies and Balashek 1952). Since then and after decades of visibly accelerating development and improvement, spoken language technology has become part of our everyday lives. Nowadays we are able to control our navigation system by voice, perform Internet voice search with our mobile phone, and navigate through complex customer self-service systems via telephone. The day our homes will become dialog partners enabled through spoken dialog technology are not too far off.

Spoken dialog systems (SDS) provide the possibility to communicate and interact with a complex computer system or application through speech, which represents the most natural means of communication between humans (McTear 2004). These systems not only simplify the interaction, but they also contribute to safety aspects in critical environments such as vehicles. For such automotive environments, hands-free operations are indispensible. That is why, SDS technology seems tailor-made to reduce the cognitive load while keeping the distraction of the driver on a minimum level. In call center environments, SDS technology—operating as *interactive voice response* (IVR) systems—helps to accomplish a variety of tasks and enables new services for users. Such systems route callers to contact persons (Riccardi et al. 1997), provide bus schedule information (Raux et al. 2005), and help out to solve technical problems (Acomb et al. 2007). They are controlled by the caller's voice. SDS can furthermore render interactions with robots and intelligent homes more natural, unobtrusive, and straightforward than a common graphical user interface (GUI) would allow. In many domains, a blend of GUI and voice user interfaces (VUI) will soon find its way into our everyday life allowing a multi-modal access to control many of our devices and applications.

Current systems predominantly follow a static and inflexible dialog flow. In order to increase the user-friendliness of an SDS and to design the dialog course in a more natural way, future systems will be able to *adapt their strategy* to the user's

A. Schmitt and W. Minker, *Towards Adaptive Spoken Dialog Systems*,
DOI: 10.1007/978-1-4614-4593-7_1,

capabilities and needs. While SDS technology in the recent years has predominantly dealt with solving problems with speech recognition, text-to-speech synthesis and basic dialog management, adaptivity still poses a major challenge to spoken dialog research.

Langley (1997) defines an adaptive user interface as an "interactive software system that improves its ability to interact with a user based on partial experience with that user". For SDS, this "experience" can be either of acoustic nature, i.e., knowledge that is derived from acoustic characteristics and the acoustic behavior of the user. Further it can be of linguistic nature, i.e., dependent on what the user says or on the user's and system's behavior affecting the entire interaction flow that is measured with interaction parameters.

A system that is aware of the interaction partner and "knows" whether it deals with a male senior novice or a young female expert user may exploit this knowledge to adapt the interaction accordingly (Metze et al. 2008). This again could happen by selecting user group-specific system prompts, which help in facilitating and accelerating the interaction, and may render systems more user-friendly. Adaptivity in SDS does not only affect static user properties, such as age, gender, expert status, etc., but also dynamic aspects affecting the interaction itself. For example, the emotional state of the user can further be a valuable source of knowledge (Cowie et al. 2001). Systems that are able to determine whether the user is angry, sad, or bored can better meet the user's needs. Interaction patterns, i.e., observable regularities in a system-user interaction, can shed light on the dialog progress and the likeliness that the task between the system and the user is completed (Langkilde et al. 1999).[1] The observation of patterns that indicate critical dialog flows may again help to adapt the strategy, and ultimately enables dialog systems to consult human assistance. This is highly interesting for SDS deployed in call-center environments in particular. Once it is detected that the dialog is unlikely to be completed, the system may escalate the call to a live agent, who can help out and finish the task jointly with the user.

In this book, we introduce the expression *online monitoring* of SDS and present innovative approaches that further advance adaptivity in such systems. The aim of online monitoring is to create knowledge for the dialog system, i.e., to derive information from "the partial experience with that user" (Langley 1997) that can be employed to render SDS more adaptive and user-friendly. While online monitoring may be introduced to acquire knowledge of arbitrary kind, we direct our attention to the statistical modeling of *critical dialog situations* that indicate malfunctioning communication. Our vision is to develop SDS that are able to react on such critical dialog situations and that adapt their strategy, once such a situation is identified. This strategy may contribute to raise task success rates, increase user satisfaction, and augment the general acceptability of SDS in the field. The *statistical and data-driven character* of the presented techniques allows a more reliable, flexible, and portable approach than hand-crafted rules could provide. Such a hand-crafted rule

[1] Interaction patterns should not be confused with "interaction *design* patterns" (Alexander et al. 1977; Gamma et al. 1995; Bezold and Minker 2011), which depict reusable templates in software engineering.

could, e.g., be: "If the automatic speech recognition engine fails to recognize the user three times in a row, then we assume that the user is dissatisfied", or "If the loudness of the user's voice exceeds a certain threshold, we assume that the user is annoyed". Such heuristic approaches rather resemble rules of thumbs, which are imprecise or even may be invalid. In contrast, this book presented approaches that are statistically backed. In a statistical method, concepts and classes (e.g., satisfied vs. dissatisfied, angry vs. non-angry) are automatically learned from large annotated training corpora and memorized in the form of model parameters. These parameters are then used by the model to determine the most likely concept or class (e.g., emotional states, dialog outcome, user satisfaction) that fits best the new observations. Relying on a statistical method with parameters determined from a large data corpus may facilitate the application of the proposed techniques to different tasks, domains, and human languages.

We consider critical situations and poor communication from three different points of view. First and foremost, it is assumed that users showing *anger* imply that they are facing a problem with the SDS. A dialog system that is able to recognize the user's emotional state could then react accordingly by changing the dialog strategy. However, as we will discover in Chap. 3, only a minority of users react emotionally when interacting with an SDS. A critical dialog situation is thus only rarely accompanied by emotions and can thus hardly be determined on the emotional state alone. Hence, we propose a more comprehensive modeling for monitoring an SDS and examine pattern-based approaches that use interaction patterns as indicators of a critical situation. Hereby, two target variables are of interest: the actual Interaction Quality and the probability of *task success*. The estimation of *user satisfaction* at arbitrary points in the interaction based on observed patterns may help future SDS to likewise adapt dialog strategies. In contrast to emotion recognition alone, the proposed approach further takes into account users that are dissatisfied without showing emotions. We finally address the issue of statistical modeling of predicting task success. Particularly for the novel generation of automated troubleshooters that help users to solve technical problems over telephone (Acomb et al. 2007) this may be of highest interest. Such automated troubleshooters help callers to jointly solve a technical problem, e.g., recover Internet connectivity. In cases where task success is unlikely, due to the course of the previous interaction, an automated troubleshooter would be enabled to escalate callers to a human operator who can help out and finally solve the problem.

The proposed techniques will allow future SDS to determine

- the emotional state of the user (the presented techniques hereby specialize on negative emotions as indicator of a critical situation);
- the user satisfaction and quality of the interaction at arbitrary points in an interaction;
- the probability that an ongoing task will reach a successful completion.

In this chapter, we describe the basic modules of an SDS. We thereby derive the need for adaptive SDS from real-life examples captured from field SDS that we present in the following. In Chap. 2, technical principles and related work are addressed and presented. We first introduce basics of machine learning and machine learning-related performance metrics, which serve as basis for the novel statistical techniques presented in this work. Afterwards, related work in speech-based emotion recognition as well as the evaluation of SDS is highlighted and discussed and shortcomings are pointed out. The newly developed and presented techniques are mainly *data-driven* and require a substantial amount of dialog interactions that have to be collected, annotated, and parameterized in order to achieve a robust statistical modeling of critical situations in human-machine communication. In Chap. 3, we thus introduce the corpora employed for developing and evaluating the developed strategies. We furthermore describe the parameterization and annotation techniques that are pivotal to all innovative strategies developed in the later chapters. Moreover, a new software platform is presented that was required for annotating and managing large corpora as well as for evaluating the statistical models.

In Chap. 4, we present novel techniques for recognizing the emotional state of the user, which we consider relevant for adaptivity in SDS. We hereby focus on the recognition of user anger and follow a holistic approach that takes into account acoustic, prosodic, and linguistic information. We further extend related approaches by introducing contextual knowledge sources. In Chap. 5, we develop novel strategies for spotting interaction patterns that indicate low Interaction Quality and thereby deduce user satisfaction. As not only the actual Interaction Quality may be relevant for spotting critical dialog situations, we direct our attention to the estimation of task success in Chap. 6. There, we apply pattern recognition with the aim of spotting dialogs that may lead to an unsuccessful outcome, i.e., failed task success. The achievements are discussed and summarized in Chap. 7 that will conclude this book.

1.1 Spoken Dialog Systems

Spoken dialog technology has sufficiently matured in recent years to allow for a broad deployment of this technology in the field. Countless applications have been evolving around the effective control of devices in automotive and mobile environments and telephone-based customer self-service in call centers (Pieraccini and Lubensky 2005). The complexity of SDS has steadily risen, which can be most clearly comprehended by looking at the development of telephone-based customer self-service applications, see Fig. 1.1.

The capability of those systems has shifted from mere information retrieval applications that have emerged in the 1990s delivering bus or train schedules toward transactional applications that enable users to manage bank accounts, book hotels, and reserve flights. The newest generation has emerged in recent years providing technical support over the telephone. The systems are able to resolve technical problems such as the recovery of lost Internet connections or the resolution of TV- and video-on-demand problems. Dialogs from such systems frequently span over 50 system-user exchanges and more posing new challenges to SDS technology.

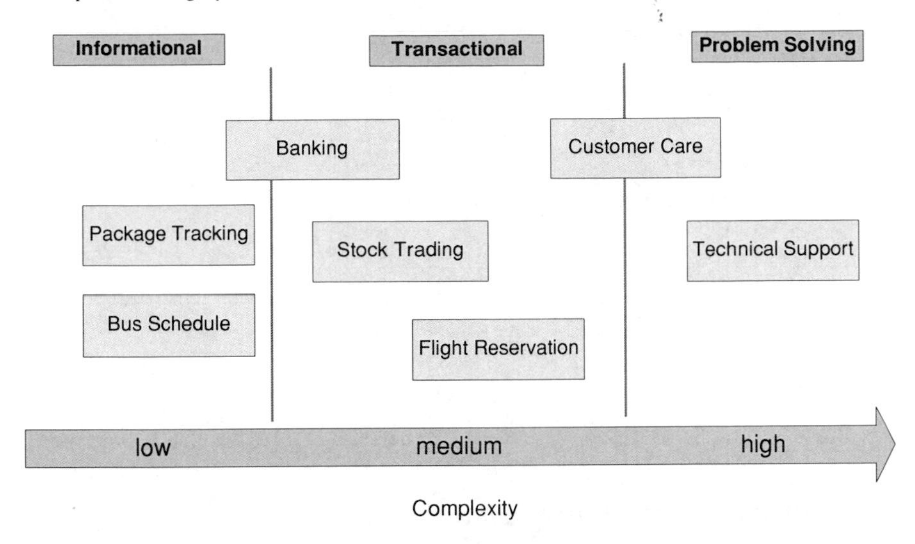

Fig. 1.1 Rising complexity of SDS in telephone-based services (Pieraccini and Huerta 2005)

While command-and-control style interaction in automotive and mobile phone settings using speech recognition still predominates, i.e., non-dialogical interaction, a trend can be observed that aims in establishing a "real" spoken dialog in smartphones and in-vehicle settings. By that, other applications than the mere voice-control of navigation systems and telematics will be enabled. Soon, we will be able to book hotels, surf the Web, and retrieve information about sights we are passing by while driving and instruct our smartphones to make doctor appointments.

However, the aim to imitate natural human–human communication with all its complexity by means of SDS technology is still far out of reach. Research on SDS has thus turned toward the development of techniques enhancing naturalness of interaction and allowing free, unconstrained user input to render spoken dialog interaction more human-like and user-friendly (Pieraccini et al. 2009). Major research activities in the SDS community thus address *adaptivity* toward specific user groups such as novice and expert users or different age groups (Kamm et al. 1998; Müller and Wasinger 2002; Müller et al. 2003; Burkhardt et al. 2007a, b; Metze et al. 2008; Schuller et al. 2011). Further research on adaptation concerns the recognition of the user's emotional state (Cowie et al. 2001; Burkhardt et al. 2005a; Pittermann et al. 2009; Schuller et al. 2010) or situation-awareness.

A SDS consists of a number of components that work together in order to enable a spoken interaction, cf. Fig. 1.2. The architecture of an SDS can be considered as a pipeline where each component processes the output of the previous one. The components are speech recognition, semantic analysis, dialog management, a language generation module, and text-to-speech synthesis (Lamel et al. 2000; McTear 2004). The modules are briefly described in the following.

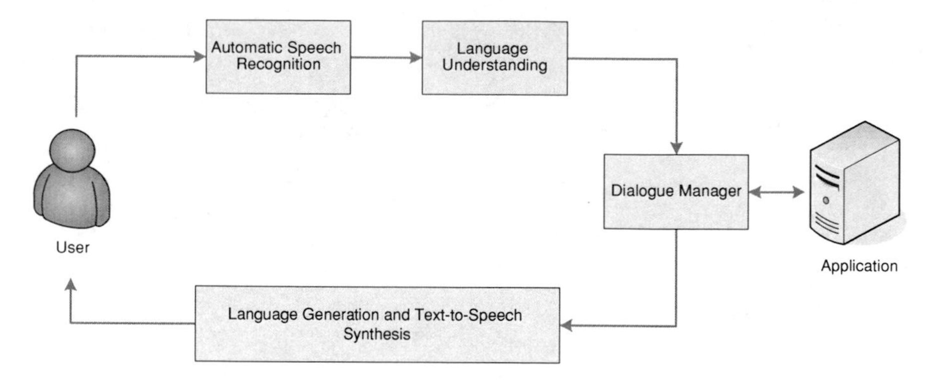

Fig. 1.2 Components of a Spoken Dialog System, similar to Lamel et al. (2000)

1.1.1 Automatic Speech Recognition

The automatic speech recognition (ASR) module captures the user's speech and converts it into a sequence of words. Formally, the goal of the ASR module is to find the most probable sequence of words $W = (w_1, w_2, ...)$ belonging to a fixed vocabulary given some set of acoustic observations $\boldsymbol{O} = (\boldsymbol{o}_1, \boldsymbol{o}_2, ..., \boldsymbol{o}_T)$. Following the Bayesian approach applied to ASR (Rabiner and Juang, 1993) the best estimate for the word sequence can be given by:

$$W^* = \arg_W \max P(W|O) = \arg_W \max \frac{P(O|W)P(W)}{P(O)}. \quad (1.1)$$

In order to generate an output the speech recognizer basically has to perform the following operations:

- extract acoustic observations (*features*) from the spoken utterance;
- estimate $P(W)$—the probability of individual word sequences to occur, regardless of acoustic observations;
- estimate $P(\mathbf{O}|W)$—the likelihood that the particular set of features originates from a certain sequence of words;
- find the word sequence that delivers the maximum of Eq. 1.1.

The term $P(W)$ is determined by the *language model*. It can be either rule based or of statistical nature. In the latter case, the probability of the word sequence is approximated through the frequencies of occurrence of individual words, often depending on the previous one or two words, in a predefined database.

The likelihoods $P(\boldsymbol{O}|W)$ are estimated in most state-of-the-art recognizers using *acoustic models* that are based on Hidden Markov Models (HMM) (see also Sect. 2.1.1). Here, every word w_j is composed of a set of acoustic units like phonemes, triphones or syllables, i.e. $w_j = (u_1 \cup u_2 \cup \ldots)$. Every unit u_k is modeled by a chain of states s_j with associated emission probability density functions $p(\boldsymbol{x}|s_j)$. These densities are usually defined by a mixture of diagonal covariance Gaussians, i.e.:

$$p(\boldsymbol{x}|s_j) = \sum_{m=1}^{M} b_{mj} N(\boldsymbol{x}, \boldsymbol{\mu}_{mj}, \boldsymbol{\Sigma}_{mj}) \tag{1.2}$$

The computation of the final likelihood $P(\boldsymbol{O}|W)$ is performed by combining the state emission likelihoods $p(\boldsymbol{o}_t|s_j)$ and state transition probabilities. The parameters of acoustic models, such as state transition probabilities, means $\boldsymbol{\mu}_{mj}$, variances $\boldsymbol{\Sigma}_{mj}$, and weights b_{mj} of Gaussian mixtures are estimated on the training stage. The total number of Gaussians being used depends on the design of the recognizer.

Finally, equipped with both $p(\boldsymbol{o}_t|s_j)$ and $P(W)$, we need an effective algorithm to explore all HMM states of all words over all word combinations. Usually, modified versions of the Viterbi algorithm are employed to determine the best word sequence in the relevant lexical tree (cf. Rabiner and Juang (1993)).

1.1.2 Semantic Analysis

While the ASR delivers a sequence of words as text, it does not return its meaning, which is a prerequisite for the dialog manager to take the next steps. The semantic analysis (SA) module (also denoted as "Semantic Parsing" or "Language Understanding") extracts from a sequence of words $W = (w_1, w_2, ...)$ this semantic meaning by using grammatical relations, rule-based semantic grammars, template matching, or statistically driven parsing techniques (Allen 1995; Jurafsky and Martin 2008). Language understanding basically comprises two processes: syntactic and semantic analysis (McTear 2004). How the words group together is determined by syntactic analysis. A common strategy to syntactically analyze a sentence is to recursively determine subphrases, such as the noun phrase and verb phrase as well as the parts of speech of the single words, such as nouns, verbs, prepositions, and modifiers. A set of rules defines a context-free grammar (CFG), e.g.:

```
1  NP-> [DET][Modifier] N [Post-Modifier]
2  NP-> NP PP
3  PP-> Prep NP
```

The rules can be interpreted as follows: a noun phrase (NP) consists of a noun (N) and optional parts in square brackets. Further, a noun phrase can consist of a noun-phrase and a prepositional phrase (PP). A prepositional phrase itself consists of a preposition (Prep) and another noun phrase. By applying such rules on a sentence, tree diagrams can be constructed representing the syntactical structure of a sentence, see Fig. 1.3.

The semantic analysis is then performed on these part of speech tags in combination with rule-based grammars. Since rule-based approaches are difficult to adapt to other domains and languages frequently data-driven methods are applied, such as Hidden Understanding Models (Miller et al. 1994, Minker et al. 1999).

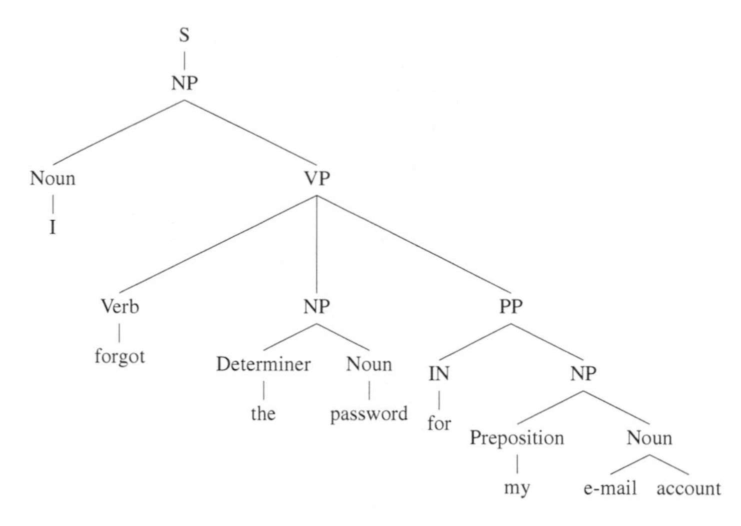

Fig. 1.3 Syntactical structure of the sentence "I forgot the password for my e-mail account"

A widely applied method in commercial SDS is to bypass syntactic analysis and to directly employ "semantic grammars". A simple semantic grammar in advanced Backus Naur form (ABNF) is depicted in the following:

```
1  $order = ($food | $drink) [please];
2
3  $food = a very* good cheese burger
4      { value="burger"};
5
6  $drink = cold+ coke
7      { value="coke" };
```

A parser using this grammar would return *food='burger'* for a sentence "A very, very good cheese burger". In a dialog system, the extracted semantic meaning is then used by the following module, the dialog manager, to choose the next action.

1.1.3 Dialog Management

The central component of the dialog system, the dialog manager (DM), is responsible for managing the dialog flow (McTear 2004; cohen et al. 2004). The dialog manager decides when and in which order it will prompt which question to the user. It accepts user input, re-prompts questions when the user does not respond in time, asks for missing information, clarifies ambiguities, and asks for confirmation when the previous components, the ASR, and LU module deliver low confidence scores.

An example dialog may proceed as follows:

```
1  0 System:     Welcome to the Bus Information.
2                What can I do for you?
3  1 User:       I want to take the bus to Bloomfield at
4                10am
5  2 System:     You want to go to Bloomfield at 10am,
6                is that right?
7  3 User:       Yes
```

In this example, the LU module delivers *departure*=‘*bloomfield*’ and *time*= ‘10:00*am*’ with a low confidence score which makes the dialog manager requesting a confirmation from the user. The dialog manager checks if all fields required to accomplish the subtask are filled. Since this is not the case, it prompts the user to specify his requirements:

```
1  5 System:     Where are you leaving?
2  6 User:       The Airport
```

The dialog manager serves as interface to the underlying application. The basic idea of this architecture is that an application might not only be interfaced by an SDS, i.e., *a VUI*, but also from a conventional desktop-based user interface, i.e., *a GUI*. A user may thus access a service either by, e.g., telephone or also by a Web front-end.

In dialog systems with information retrieval and transactional character the application may contact a database (e.g., to search for bus schedules) or perform a transaction (e.g., to book a hotel) (Cohen et al., 2004). In problem-solving applications, such as an automated Internet troubleshooter, the task may be a different one. The application may, e.g., reset the caller's modem, send a ping signal, or find out whether an outage in the neighboring network occurred.

Many different approaches exist to implement DMs. The dialog manager is often realized as finite-state machine that follows a hand-crafted rule set to choose the next action, such as the W3C standardized VoiceXML description language (Oshry et al. 2007). It provides a model description of the dialog defining the structure of a specific dialog. Other rule-based frameworks such as the TrindiKit (Larsson and Traum 2000) and agent-based systems, Olympus (Bohus et al. 2007), exist. They need strong presumptions about the implementation scenario leading to an inflexible and constrained system. Examples for statistical approaches are the Bayes Net Prototype implemented within the TALK Project (Young et al. 2006), or dialog managers based on Partially Observable Markov Decision Processes (POMDP) (Williams and Young 2007). Another recent approach incorporates ontologies that model the dialog flow in an SDS, see, e.g., Heinroth et al. (2010a).

1.1.4 Language Generation and Text-to-Speech Synthesis

The last two modules, text generation and text-to-speech (TTS) synthesis, are closely tied together and cannot always be considered as separate parts. In cases where the messages are static, prerecorded prompts from professional speakers are presented to the user. Most commercial systems use such "canned speech" (Möller 2005b; McTear 2004).

In general, response generation is concerned with decisions on which information is provided to the user and in which structure and form (words and syntax) it is presented. Ideally, text generation involves user models that take into account the user profile and current user state (Paris 1988). A simple, yet inflexible approach, is the use of canned text. It is only useful if the information contained represents the requested information. Template filling offers more flexibility and allows to slightly varying the prompt. Database information can be parsed into the template, e.g., "The train to *$destination* leaves at *$dep_time* from platform *$platform_no*.", where *$destination*, *$dep_time* and *$platform_no* represent results from a database query. Template filling is a common approach in VoiceXML, cf. Larson (2002). More advanced methods view text generation as planning process that starts with a communicative goal and returns a text message (Reiter and Dale 2000).

The generated text is brought into spoken form by the speech output module. The output is either completely synthesized, constructed with prerecorded, canned prompts or it constitutes a mixture of both. A text template as in

```
So you want a \$room_type from the \$arr_date to the
\$dep_date?
```

would ideally be brought into spoken form by canned prompts where the variables *$room_type*, *$arr_date* and *$dep_date* are either pre-recorded parts or generated with a Text-To-Speech synthesis module (McTear 2004). A mixture between canned and synthesized parts may often leave an unnatural and patchy impression to the listener.

TTS synthesis converts a text to a sequence of phonemes or diphones by using a dictionary. Naturalness of the voice is added by performing a phoneme to speech conversion. It adds prosody, comprising rhythm, intonation, loudness, and tempo to the linguistic description (McTear 2004). The modeling of prosody is considered major challenge in TTS. A poor prosody modeling results in unnatural sounding voices.

1.2 Towards Adaptive Spoken Dialog Systems

Let us assume a customer of a railroad company comes up to a counter to purchase a railroad ticket. The clerk enquires about the departure and the destination, the time of departure and further details required to issue the ticket. In this scenario, a professional clerk would adapt the interaction according to the customer, for which he may use a large number of information sources, such as:

- social information about the user, e.g., estimated personality, age, educational background, gender, etc.;
- the emotional state of the user, derived from facial expressions together with acoustic and lexical information;
- discourse information, i.e., the verbal information exchanged between clerk and customer;

- the probability of task success, i.e., how likely the concern of the customer can be handled;
- context information, e.g., the standing time the customer had to wait in line;
- environmental factors, such as the noise level in the ticket hall or loudspeaker announcements;
- the user's experience with the task.

Based on these properties adhering to the counterpart and the environment, the clerk may adapt his way of interacting. For example, he might choose a less formal speaking style when speaking to a young customer, he might increase his voice if the customer shows signs of a hearing deficiency, or might switch to a different language when noticing that the customer is a foreign guest.

Among many other fields of application, SDS are increasingly being deployed to serve users for such a task over telephone or kiosk systems. However, the restriction applies that modern SDS are incapable of adapting as the clerk does in the example. Instead modern SDS are mainly static in terms of what they prompt to the user and static in terms of how they treat the user notwithstanding the user-specific properties and the course of the previous conversation. The rising complexity of SDS demands for innovative techniques that help rendering future systems interaction aware for enabling adaptivity. Ultimately, this will lead to more natural interactions, higher acceptance, and raised usability of SDS. With the knowledge gained during the interaction a dialog system would be capable of adapting its strategy, similarly as a real, customer-friendly clerk would do.

In SDS this adaptivity can be reached by two basic steps: a *detection* step and an *action* step. Both steps may, respectively, be arranged in the form of a wheel as the central terms "detection" and "action" consist of several subcategories, which are explained with concrete examples placed at the outer edge of the wheel. These *Adaptivity Wheels* are depicted in Fig. 1.4.

Before adaptivity may happen, an SDS needs to gain knowledge about the user and the interaction, which is shown in the Detection Wheel, see Fig. 1.4a. Based on observations made during an interaction, the system may estimate static and dynamic user properties, as well as properties adhering to the interaction itself. With the help of trained data-driven, stochastic models, specific target variables may be predicted, which are part of *static user properties*, e.g., the gender and age of the user, or in terms of *dynamic user properties* the emotional state, an intoxication level or the user's current interest level. Moreover, the *interaction-related properties* may be determined by evaluating the interaction itself, which may for example reveal that the user seems dissatisfied with the current interaction, or that the ASR performance delivers poor results.

Following the detection phase, a certain action can be triggered, see the *Action Wheel* in Fig. 1.4b. This is where the actual adaptivity happens based on the gained knowledge. As for *speech input*, an adaptivity action might be to switch standard acoustic models to gender- or accent-specific models, which have proven to show better results in ASR. The *speech output* may be adapted with gender information, such as selecting a male voice instead of a female voice, attracting specific user

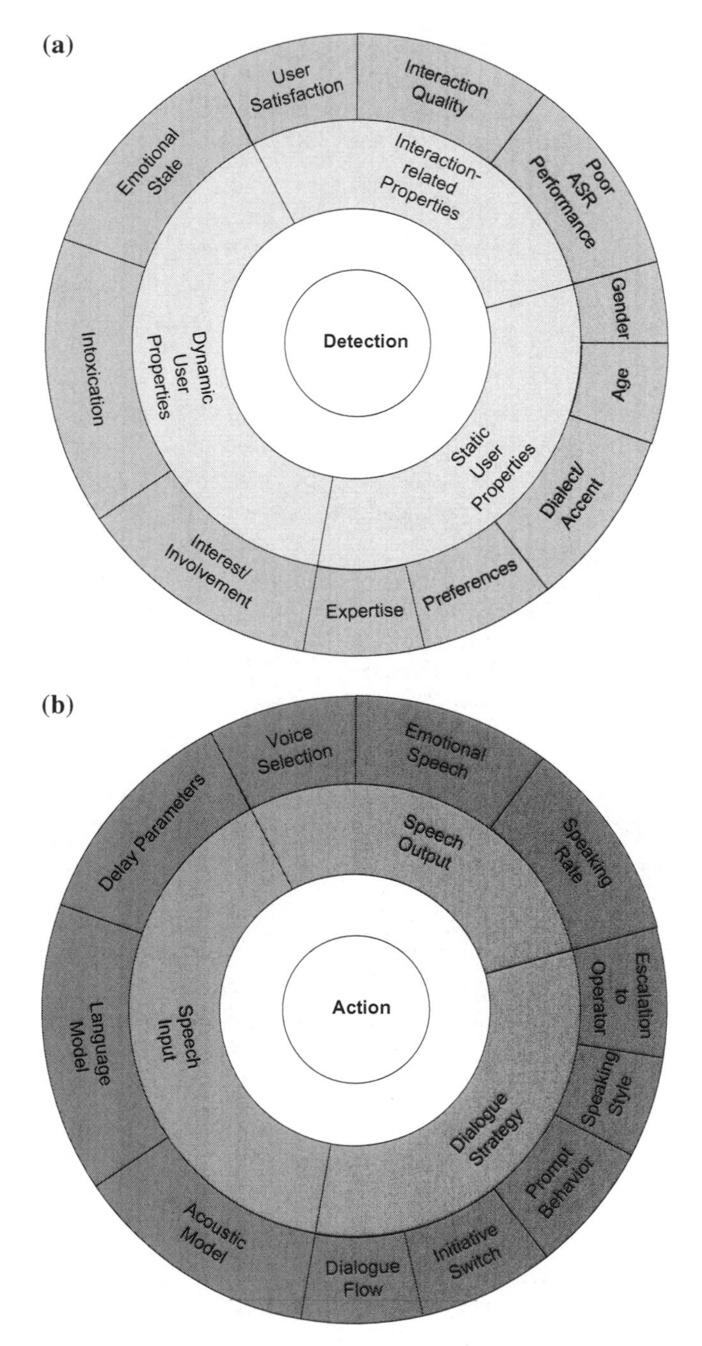

Fig. 1.4 Adaptivity wheels. **a** Detection wheel: on what to adapt? **b** Action wheel: how to adapt?

groups. Likewise, the *dialog strategy* may be changed. Having estimated poor ASR performance, the system may switch to more direct confirmations, choose different and more explanatory system prompts or ultimately escalate the task to professional human assistance, e.g. in telephone-based systems. Obviously, the question what has to be done with specific information obtained from the detection phase cannot be clearly defined and depends on the domain and the requirements of a system.

Today's SDS specifically face the problem that speech recognition and language understanding are error-prone and both are—with very few exceptions—unable to cover free user input. This frequently leads to critical dialog situations where the interaction between the system and the user ends in failure and where task success is not achieved. Miscommunication due to poor system performance and design but likewise false user behavior and wrong user expectations may leave behind unsatisfied users. In the worst case, this even leads to an abortion of the task. To illustrate problems that may arise during a spoken human-machine interaction and that may be solved through adaptivity, two examples are described in the following, both originating from system logs of commercial and non-commercial telephone-based real-life SDS. The first example dialog shows some resemblance to the customer-clerk scenario as it originates from a deployed bus information system for the city of Pittsburgh, serving callers with bus schedules. The transcript of the interaction is depicted in Table 1.1.[2] In this example, speech recognition and understanding errors are pieces of evidence for communication going badly. The user reacts angrily and finally ends the call early without finishing the task. Attention should be paid to the emotional state that was annotated by three independent human expert raters for this dialog.

Similar situations can be observed in other domains. An excerpt of a critical dialog between a user and an automated Internet troubleshooter (cf. Chap. 3) is depicted in Table 1.2.

In this automated technical support scenario, the system and the user try to repair a lost Internet connection. Here, the user pretends that he has no modem, while the system insists in asking for the modem's brand. Certainly, the SDS is working correctly at this moment since the discussed cable modem naturally belongs to the standard equipment of each customer. However, situations like these could be prevented if the SDS was capable of detecting the caller's frustration and recognizing that the dialog gets out of step. Several indications of a critical setting are visible: the system mistakes the user utterances and consecutively demands the very same information; the user frequently barges in and interrupts the system prompt; the user expresses his displeasure by swearing and shouting. Again, a certain pattern is visible that indicates a critical dialog situation.

Most urgently, future dialog systems require a solution that permits monitoring of the interaction between the system and the user and that estimates the quality of an interaction as well as the probability of task success covering the emotional state and interaction patterns. The benefits would be twofold:

---

[2] Later we will use more data from this domain for the development of statistical prediction models. We will denote this corpus as LG-FIELD.

Table 1.1 Example dialog of a malfunctioning interaction between a caller and the CMU Let's Go bus information system (Raux et al. 2006)

System:	Welcome to the CMU Let's Go bus information system. To get help at any time, just say Help or press zero. What can I do for you?	
User:	54C schedule *[54C]*	NonAngry
System:	The 54C . Did I get that right?	
User:	Yes. *[yes]*	NonAngry
System:	Alright. Where do you want to leave from?	
User:	West North Avenue *[/noise/ /mumble/]*	NonAngry
System:	Leaving from EIGHTH AVENUE . Did I get that right?	
User:	No, West North Avenue. *[eighth_avenue]*	SlightlyAngry
System:	Leaving from EIGHTH AVENUE . Did I get that right?	
User:	No! *[north_avenue]*	Angry
System:	Leaving from NORTH AVENUE . Is this correct?	
User:	Arch Street and Cedar Avenue *[no]*	Angry
System:	Leaving from FIFTH AVENUE . Did I get that right?	
User:	what?! *[/noise/]*	Angry
System:	Okay. Where are you going?	
User:	I'm not leaving from fifth_avenue! *[/noise/]*	SlightlyAngry
System:	Going to EIGHTH AVENUE . Is this correct? <User hangs up.>	

The place of departure is not correctly recognized by the ASR module. Noise, the wrongly parsed user answers and the user's emotional state indicate that task success is unlikely to reach. There is evidence that the user is unsatisfied with the situation given the acoustic impression

- Long lasting dialogs could be avoided if the system itself detects early enough that it cannot be successful in the given task, when estimating that the user satisfaction is low due to the previous interaction or when detecting user anger. In that case, it might perform an early escalation to a live agent who will solve the task. By that, the user does not uselessly spend time with automation when the task presumably will not be completed.
- Moreover, an adaption of the dialog strategy that is triggered when critical situations are automatically detected could prevent task failures and prevent users from a premature end of the task. This adaption could be accomplished e.g. by introducing more direct confirmations or by adding domain-dependent steps, which could result in more natural and user-friendly interactions.

It is obvious that an early detection of troublesome communication may have several benefits for the SDS and the user. To achieve this we propose an additional module that allows monitoring ongoing system-user interactions, see Fig. 1.5.

The suggested online monitoring module (OMM) observes the ongoing interaction based on captured log data and user recordings provided by the ASR, LU and DM modules. Interaction patterns, static and dynamic user characteristics are recognized based on stochastic models and the generated knowledge is passed as *knowledge vector* to the dialog manager, which may react on the observations (i.e., trigger an *action*). Each vector component thereby originates from an individual subcomponent

Table 1.2 Example dialog from the LevelOne Broadband Agent demonstrating a typical emotional situation caused by misunderstandings: The system asks the caller to identify the modem and its brand name

	...	
System:	So you have a Netgear modem, is that right?	
User:	yes *[yes]*	NONANGRY
System:	Before we go on, I need you to make sure that Netgear has a light on the front labeled 'Status.' When you've found out whether or not your Netgear has a Status light, say Continue.	
User:	continue *[continue]*	NONANGRY
System:	[...] So, is there a Status light on the modem?	
User:	no *[no]*	NONANGRY
System:	Well, if it doesn't have that Status light, then I think you're looking at your Netgear *router*. [...] When you've found your modem, say Continue.	
User:	_noise_ *[_nomatch_]*	
System:	Please, just say Continue after you've found your modem.	
User:	continue *[continue]*	NONANGRY
System:	What's the brand name?	
User:	_noise_ *[_nomatch_]*	
System:	I didn't quite get that. Please tell me the brand of your modem again, or, if there is no brand printed on it, say No Brand Name.	
User:	no modem *[no_brandname]*	SLIGHTLYANGRY
System:	So your modem does *not* have a brand name printed on it, right?	
User:	there is no modem *[_nomatch_]*	SLIGHTLYANGRY
System:	So your modem does *not* have a brand name printed on it, right? Please say Yes or No.	
User:	no modem *[_nomatch_]*	ANGRY
System:	So your modem does *not* have a brand name printed on it, right? Please say Yes or No.	
User:	no *[_nomatch_]*	ANGRY
System:	Let's just move on for now.	
System:	Now that we've correctly identified the modem, I need you to unplug it. When it's unplugged, say Continue.	
User:	no, there is no * modem! give me a * person! *[_nomatch_]*	ANGRY
	...	

The caller is pretending that there is no modem, a case the system does not intercept since the absence of a modem would prevent an Internet connection. Both, the system and the user are unable to cope with this situation and the caller is getting angry. The caller's answers comprise hand transcriptions, natural language parse (in italic) and emotional state

that encapsulates the respective statistical model. We will denote these models also as *recognizers*, since their task is to recognize a specific user characteristic or interaction situation. The knowledge derived from the observed interaction is thereby not strictly defined and predominantly depends on the requirements of the application. In this work, we propose submodules that contribute to the resolution of critical situations, since they are most likely to occur in SDS. Our proposed components recognize the emotional state of the user (EMO), estimate user satisfaction (US) as

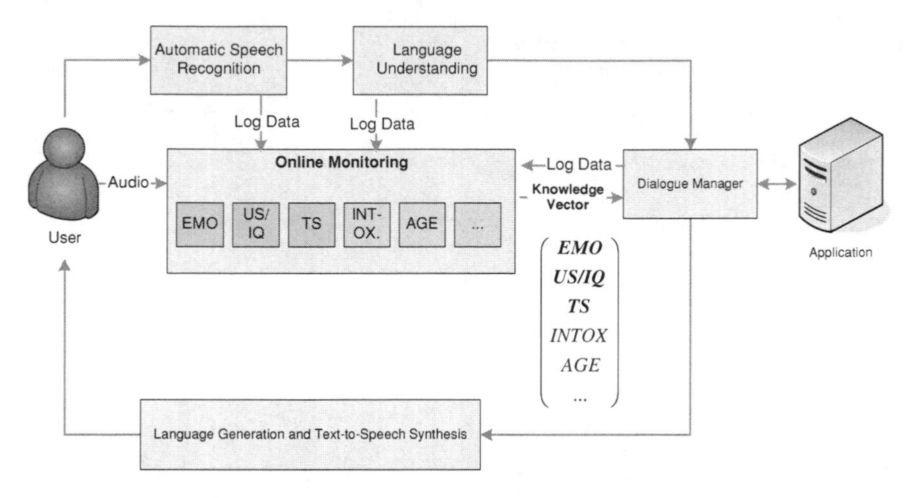

Fig. 1.5 Enhanced, adaptive SDS with Online Monitoring Module. The OMM encapsulates a number of submodules that derive knowledge from the ongoing interaction. Each submodule itself encapsulates a statistical model/recognizer. It delivers knowledge to the DM that allows it to render the interaction more adaptive and user friendly. Depicted are the submodules to recognize the emotional state (EMO), user satisfaction/Interaction Quality (US/IQ) and task success (TS). Further subcomponents are depicted, such as intoxication (INTOX.) and speaker age (AGE), which are not further discussed in the course of this book

well as interaction quality (IQ), and determine the probability of task success (TS). Other subcomponents, that are not developed in this work, but fit into the proposed architecture could further contribute to adaptivity in SDS. In principle, it can be noted that there is no silver bullet when it comes to the definition of submodules. Basically, any knowledge that can be derived from log and audio data could enhance the OMM as long as the specific knowledge generated by the submodule is of help for the application. For example, an intoxication submodule that determines whether the speaker is drunk by analyzing the voice may be of particular interest in online monitoring automotive SDS.[3] Other subcomponents that determine gender, dialect, or non-native speakers may help to adapt acoustic models for improving speech recognition or adapt the system's prompting strategy. Likewise, the estimation of speaker age might lead to an adapted dialog that satisfies the needs of specific age groups.[4] All topics addressed in this book focus on the detection part in adaptivity.

[3] Our approach to accomplish this is described in Ultes et al. (2011b), but is not further discussed at this point.

[4] Age recognition is discussed in Schmitt et al. (2010d).

Chapter 2
Background and Related Research

This book presents novel techniques for modeling spoken human-machine interaction to foster adaptivity in SDS. Primarily we are interested in the detection of critical dialog situations whose symptoms are certain emotions as well as particular observations made during an interaction. Both, emotions and these interaction observations are "patterns" that may be statistically modeled. Therefore, pattern recognition and machine learning (ML) approaches are pivotal to the following chapters. The presented approaches consequently do not rely on static rules, but may be derived and learned from data and may thus be easily ported to systems operating in new domains.

This chapter provides the technical and theoretical background required for the following chapters. Additionally, related work is introduced, discussed and shortcomings are pointed out. In Sect. 2.1, an introduction to the basics of supervised ML is given and evaluation metrics that are applied to assess the presented classifiers and techniques are presented.

Subsequent to this, the theoretical background for recognizing emotions from speech is provided in Sect. 2.2. Here the initially introduced ML techniques are of help. Basic emotion theories, existing speech-based emotion corpora, common techniques, relevant paralinguistic and linguistic features as well as related work and common shortcomings are presented.

Negative emotions are only one symptom for a poor "communication" between a system and a user. Certain interaction patterns may indicate that something is going wrong, even if no emotions may be observed. In Sect. 2.3, we thus introduce pattern-based techniques that allow the estimation of quality-related target variables in the field of SDS. We begin with approaches that aim to estimate *overall system quality* based on observed interaction patterns. These approaches are designed to work *offline*, i.e. the quality score is estimated on *completed dialogs*. The score is used to quantify the usability of a system. Afterwards, we present related work that may be used to assess quality *online*, i.e. *during* the interaction. We hereby describe approaches that estimate user satisfaction and aim to predict task success.

A. Schmitt and W. Minker, *Towards Adaptive Spoken Dialog Systems*,
DOI: 10.1007/978-1-4614-4593-7_2,

2.1 Machine Learning: Algorithms and Performance Metrics

Discriminating between the written letters A and B, determining a ripe apple by smelling or distinguishing between Beethoven and the Beatles by listening are tasks that we humans accomplish with ease. The field of ML has been trying to immitate this human capability and meanwhile computers are able to recognize faces, sort out dirty bottles at assembly lines and tell us, which radio song we are listening to when we record a short sample with our mobile phone. The fields of application of ML are very diverse and allow computer vision, speech recognition and other arbitrary data-driven learning processes, such as the online monitoring techniques that have been implemented in this book.

Basically, two main branches can be discriminated: supervised and unsupervised learning. While some special forms of ML do exist, they are not further addressed here, such as reinforcement learning and semi-supervised learning. In ML, objects are represented as a set of features that characterize the object. Formally speaking, an object o is described as feature vector

$$f = \begin{pmatrix} f_1 \\ .. \\ f_n \end{pmatrix}. \quad (2.1)$$

Supervised Learning requires that it is known beforehand to which class c_i an object o of a training set belongs to. This class membership is also referred to as "label". The classifier is trained on a set of objects from all possible classes (the training data) with an arbitrary pattern recognition algorithm. Examples of these algorithms are presented later in this chapter. The algorithm infers a function that is called discriminative classifier if the label of the object is discrete. It is then able to assign a discrete class to an unseen object (*classification*). If the label is not a discrete class but a continuous value, a regressive supervised learning algorithm infers a regression function that can be used to approximate the continuous target value (*regression*).

Unsupervised Learning assumes in contrast to supervised learning that the class memberships of the objects are unknown. The task is then to find a hidden structure among these objects, i.e. possible classes, for example by applying clustering algorithms, such as the k-means algorithm. Clustering divides the data into k different groups (the clusters) where the objects of one cluster share a specific property. To illustrate this, imagine we are trying to cluster speech samples from male and female voices that are represented by a single average pitch value. Let us further assume that the label of each sample is not known to us. Given the premise that we want the unsupervised learning algorithm to separate the samples into two classes (and by that automatically label them) the algorithm splits the samples into two disjoint groups (the *clusters*). After successful clustering, one cluster predominantly consists of "male" and the other cluster of "female" samples. The clustering algorithm can make use of the fact that male voices usually exhibit a lower pitch than female voices.

Unsupervised learning is not relevant for solving problems addressed in this book and is thus not discussed any further. For a deeper insight into unsupervised learning refer to Albalate and Minker (2011).

2.1.1 Supervised Learning

In the following, we introduce the topic of classification with a rather simple learning algorithm, the *k-Nearest Neighbor* (KNN), before we explore more advanced algorithms that are predominantly employed in speech- and language processing. Hereby we will outline Support Vector Machines and Hidden Markov Models (HMMs). Finally, we address Linear Regression as a representative for function approximation that can be used for predicting continuous, numeric labels.

k-Nearest Neighbor

The nearest neighbor classifier is a representative of instance-based learning algorithms, since it stores the training samples verbatim and uses a distance function to determine the closest training sample to the test sample (Witten and Frank 2005). To determine the class of an unknown sample σ, the nearest neighbor algorithm determines the distance to all data points in the training set and assigns the class of the closest data point to σ. Usually, the Euclidean distance is applied in case the features depict continuous variables,

$$d(r, h) = \sqrt{(r_1 - h_1)^2 + (r_2 - h_2)^2 + \cdots + (r_n - h_n)^2} = \sqrt{\sum_{i=1}^{n}(r_i - h_i)^2}, \tag{2.2}$$

where r is the reference vector of sample σ and h a hypothesis vector from the training set. Depending on the task, any other metric may be applicable (Schürmann 1996). More robust results can be achieved by taking into account the k closest data points (k-nearest neighbor) and by performing majority voting to determine the final class. Figure 2.1 depicts a k-nearest neighbor classifier (here $k = 3$) for classifying two-dimensional data objects in a three-class classification problem. Let us assume the objects represent fruits and the classifier's task is to classify an unknown fruit to one of the classes "grapes", "plums" and "apples". Each fruit is represented by a two-dimensional feature vector (x, y) where x represents the weight and y the diameter of the fruit. To assign the new object to one of the classes, the distance to the three closest data points is relevant. In this example, the three closest data points are a (=grape) as well as b and c (=plums). According to the majority vote the k-nearest neighbor classifier would select "plum" as final class for σ.

Nearest neighbor classifiers are also called *lazy learners*, since there is no obvious learning process involved (Witten and Frank 2005). The actual work is done at the

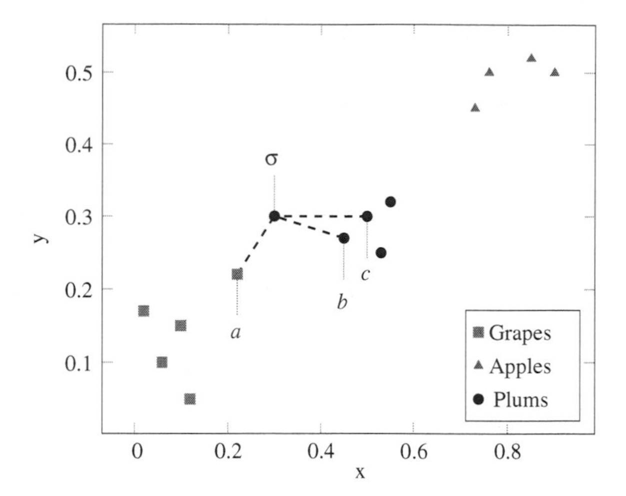

Fig. 2.1 Illustration of k-nearest neighbor classification (here $k = 3$) in a three class recognition task

time of classification, and it is easy to comprehend that a large computational effort is required at that point. The complexity for classifying an object hereby rises with increasing size of the training set and the application of k-nearest neighbor can be extremely time-consuming and memory-intensive (Webb 2002).

Support Vector Machine

One of the most powerful and popular classifiers are Support Vector Machines (Cortes and Vapnik 1995) (see also Bennett and Campbell (2000)) since they yield high predictive performance and are fast during classification. Furthermore, SVMs are robust towards overfitting to training data. Overfitting occurs when a learned model better memorizes the characteristics of the training data than being able to generalize on unseen test data. SVMs are able to robustly solve both linear and non-linear problems.

Based on a set of training examples

$$\{(\mathbf{f}_1, c_1), \ldots, (\mathbf{f}_m, c_m) | \mathbf{f}_i \in \mathscr{F}, c_i \in \{-1, 1\}\} \tag{2.3}$$

an SVM determines a hyperplane that separates the training sample of two classes $\{-1, 1\}$ in such a way that the margin between classes is maximized. In Fig. 2.2 such a hyperplane is shown for a two class data set. The instances that are closest to the hyperplane are called *support vectors*, the encircled points in the figure. They are equally close to the hyperplane and are thus the most difficult to classify. These support vectors define the maximum margin hyperplane x

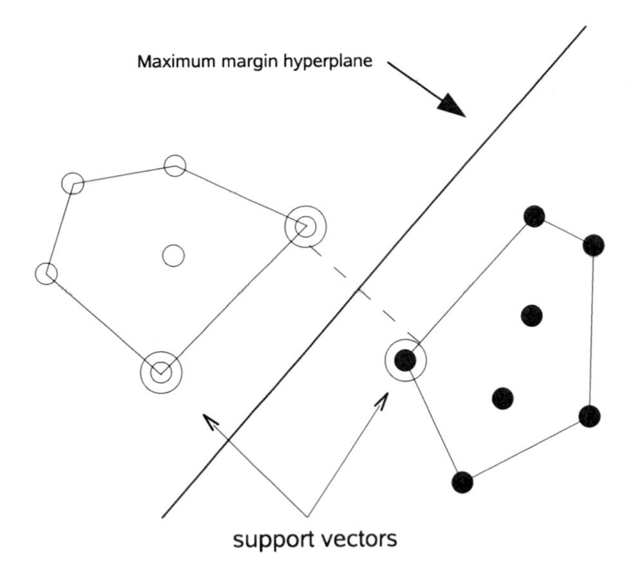

Fig. 2.2 Maximum margin hyperplane and support vectors (Witten and Frank 2005)

$$x = b + \sum_{i=1}^{n} \alpha_i y_i \mathbf{a}(i) \cdot \mathbf{a} \tag{2.4}$$

where $y_i \in \{-1, 1\}$ is the class label of the training instance $\mathbf{a}(i)$, b and α_i are numeric parameters describing the hyperplane that have to be determined by the learning algorithm. The term $\mathbf{a}(i) \cdot \mathbf{a}$ is the dot product of the test instance $\mathbf{a}$ with one of the support vectors $\mathbf{a}(i)$. Generally speaking it can be said that the larger the margin, the better the generalization of the SVM. A single SVM can solve only binary classification problems. Multi-class problems are usually solved by dividing the classification in several "one-vs-all" classifiers.

Special to SVMs in contrast to other learning algorithms is that they choose the hyperplane with the maximum margin to separate the classes. For non-linear problems, i.e. classes where the samples cannot be separated by a linear hyperplane, the kernel trick (Scholkopf and Smola 2001) is applied.

Linear Regression

Linear regression can be applied for supervised learning tasks that require a numeric prediction, i.e. the label of the samples is a continuous integer or real-valued number (Witten and Frank 2005). Regression aims to find a functional description of data that can be used to predict values of new, unseen input (Duda et al. 2001). It thus belongs to the function approximation approaches (Schuermann 1996). We will see in Sect. 2.3.1 that a linear regression model can be used to predict user satisfaction

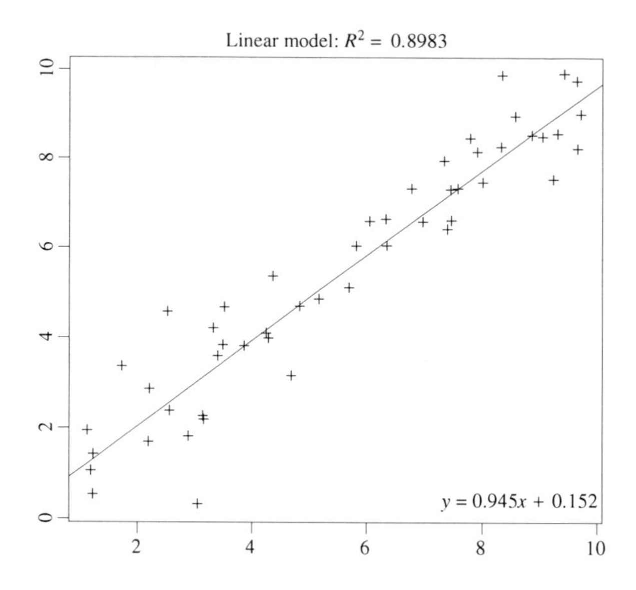

Fig. 2.3 Example for a linear regression. The random process is approximated using a linear function that minimizes the error

in an SDS based on interaction data from a set of dialogs. In linear regression the prediction is expressed by a linear combination of the input features with weights that are obtained during the training iterations. Figure 2.3 depicts a linear regression model.

A general linear regression function is described as follows:

$$y = w_0 + w_1 f_1 + w_2 f_2 + \cdots + w_n f_n \tag{2.5}$$

where w_i are the weights obtained by the training process and f_i the single attribute values of the unseen sample. The selected weights are those that fulfill the *least mean-square* criterion, which reaches its optimum when the sum

$$S = \sum_{i=1}^{n} r_i^2 \tag{2.6}$$

of squared residuals r_i is minimum. Hereby, a residual is the difference of the actual value to the predicted value of the linear regression function (Schuermann 1996).

Hidden Markov Models

In contrast to the previous approaches, where an object is modeled as single feature vector, HMMs can be employed for recognizing objects of sequential or temporally changing nature, i.e. where the object can be represented as sequence of feature vectors. HMMs have proven to be superior for example in speech and gesture recognition.

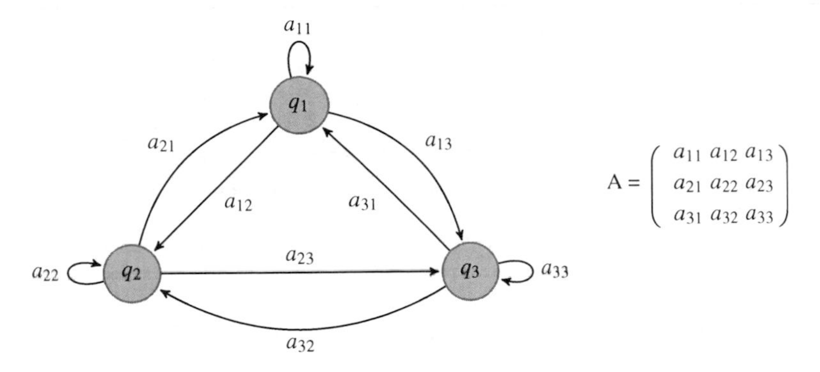

Fig. 2.4 Markov chain with three states

In speech recognition, HMMs model the dynamics of speech as sequences of feature vectors where each vector represents a short time frame of the speech sample. This yields much more accurate results compared to template matching approaches e.g. with k-Nearest-Neighbor that considers the entire speech sample as a single feature vector consisting of mean values (Pfister and Kaufmann 2008).

The simplest and most often used form of Markov models are the *Markov chain models*, which are discrete Markov models whose observations o_i are uniquely corresponding to the states q_i of the model. Higher-order Markov chains offer principal advantages over first-order chains, since the longer temporal dependency can always be coded into a single state by an appropriate extension of the state space. In any discrete first-order Markov model the probability distribution of a first order Markov chain can be written as:

$$P(q_t \mid q_1, q_2, q_3, \ldots, q_{t-1}) = P(q_t \mid q_{t-1}). \tag{2.7}$$

Containing a finite number of states m, the conditional probabilities for the transition to state j under the condition of being in state i can be compactly written into a matrix of *state transition probabilities*, which completely describes the property of the Markov chain:

$$A = \{a_{ij}\} = \{P(Q_t = j \mid Q_{t-1} = i)\} = \begin{pmatrix} a_{11} & \cdots & a_{1m} \\ \vdots & \ddots & \vdots \\ a_{m1} & \ldots & a_{mm} \end{pmatrix}, \quad \sum_j a_{ij} = 1 \, \forall i. \tag{2.8}$$

A first-order Markov chain with a finite number of states can be illustrated as a finite state machine as depicted in Fig. 2.4.

Markov chains are widely used for different applications in fields such as queueing theory, physics, mathematical biology (population processes) or Internet-related algorithms (spam filters, page ranking, personalized contents).

Extending the described Markov chain with a finite state space by adding a second stage stochastic process yields a *Hidden Markov Model* (HMM). An underlying Markov chain now produces an output that is no longer unique to the current state of the model. It is random, according to a probability distribution associated to the single states of the chain. Since Markov models by definition produce only sequences of outputs the chain as the basic model now is "hidden" when observing its behavior, from which the term *hidden* Markov models is derived.

The properties of a first order HMM λ are completely described by:

- The finite number of states $\mathbf{N}$.
- The state transition probability distribution matrix $\mathbf{A}$, see Eq. 2.8.
- The state-specific output or observation probability distributions vector $\boldsymbol{B}$

$$B = \{b_j(\mathbf{x})\} = \{P(\mathbf{x} \mid Q_t = j)\} = \begin{pmatrix} b_1(\mathbf{x}) \\ \vdots \\ b_N(\mathbf{x}) \end{pmatrix}, \quad \int b_i(\mathbf{x})\, \mathrm{d}\mathbf{x} = 1 \forall i, \mathbf{x} \in \mathbb{R}^n. \tag{2.9}$$

- The initial state probabilities vector π

$$\pi = \{\pi_i \mid \pi_i = P(Q_1 = i)\} = \begin{pmatrix} \pi_1 \\ \vdots \\ \pi_N \end{pmatrix}, \quad \sum_i \pi_i = 1. \tag{2.10}$$

The tuple $\lambda = (A, B, \pi)$ will be used as a compact notation to denote an HMM (Rabiner and Juang 1993).

It should be noted that the output probability distribution B of the model can be either of discrete or continuous character, which discriminates *discrete HMMs* from *continuous HMMs*. In the following, we will consider only discrete HMMs and refer to Rabiner (1990) for further reading.

Discrete HMMs exhibit discrete output probability distributions. Every state i of the model emits an observation out of a finite set of m observation symbols $v_i \in V$ according to an attached probability vector containing the emission probabilities of the distinct symbols for this state. So the output distribution B of a discrete HMM can be written as a matrix:

$$B = \{b_{im}\} = \begin{pmatrix} b_{11} & \cdots & b_{1m} \\ \vdots & \ddots & \vdots \\ b_{i1} & \ldots & b_{im} \end{pmatrix}, \quad \sum_m b_{im} = 1 \, \forall i. \tag{2.11}$$

In relevant literature an introductory example for HMMs normally is given by the use of a discrete HMM, see Fig. 2.5.

The figure describes an HMM of a weather model. Let us assume a prisoner in jail without daylight wants to estimate the weather outside. He knows that if there is a sunny day the next day will be likewise sunny with a probability of 90 % and

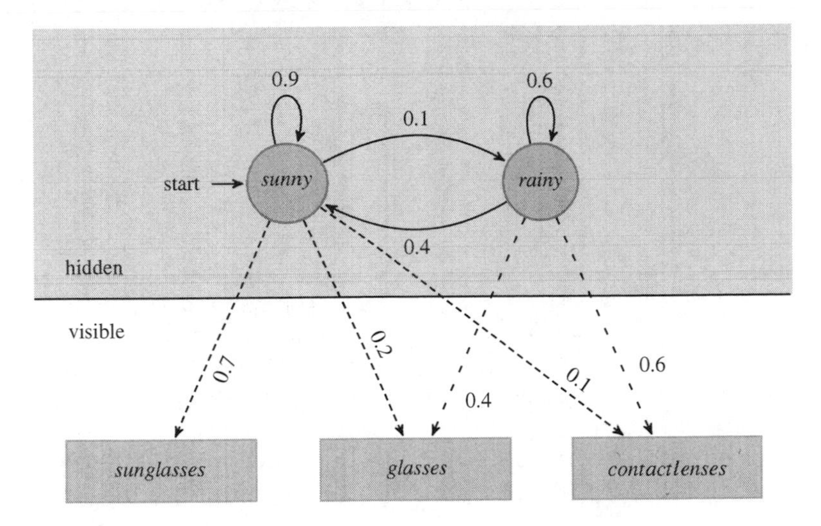

Fig. 2.5 Example for a Hidden Markov Model

rainy with a probability of 10 %. Furthermore, if there is a rainy day the following day will also be rainy with 60 % probability and sunny with 40 % probability. Finally he, knows that the day he was brought to jail the weather was sunny.

His second information source is the eyewear of the jail guard that enters the jail every day to provide food. The prisoner knows that on sunny days the jail guard wears sunglasses with a chance of 70 %, normal glasses with the chance of 20 % and contact lenses with a 10 % probability. On the other hand, on rainy days he usually wears normal glasses in 40 % of all cases or contact lenses in the remaining 60 %.

The prisoner can model an HMM with all of this information. By observing the eyewear of the jail guard he can now draw conclusions on the "hidden" weather outside. The HMM $\lambda = (A, B, \pi)$ in this case is defined by

$$A = \begin{pmatrix} 0.9 & 0.1 \\ 0.4 & 0.6 \end{pmatrix}; \quad B = \begin{pmatrix} 0.7 & 0.2 & 0.1 \\ 0.0 & 0.4 & 0.6 \end{pmatrix}; \quad \pi = \begin{pmatrix} 1.0 \\ 0.0 \end{pmatrix} \tag{2.12}$$

along with the discrete set of states $Q = \{\text{sunny, rainy}\}$ and the discrete set of distinct observation symbols $V = \{\text{sunglasses, glasses, contact lenses}\}$. With that model and watching the jail guard wearing sunglasses on the first and contact lenses on the two following days he can calculate the probabilities

$$P(O, Q = \{q_1, q_2, q_3\} \mid \lambda) = \pi_{q_1} \cdot b_{o_1} \cdot a_{q_1 q_2} \cdot b_{o_2} \cdot a_{q_2 q_3} \cdot b_{o_3} \tag{2.13}$$

for the observation sequence

$$O = \{\text{sunglasses, contact lenses, contact lenses}\} \tag{2.14}$$

for all possible weather conditions:

$$\begin{aligned}
P(O, Q = \{sss\} \mid \lambda) &= 1.00 \cdot 0.70 \cdot 0.90 \cdot 0.1 \cdot 0.9 \cdot 0.1 = 0.00567 \\
P(O, Q = \{ssr\} \mid \lambda) &= 1.00 \cdot 0.70 \cdot 0.90 \cdot 0.1 \cdot 0.1 \cdot 0.6 = 0.00378 \\
P(O, Q = \{srs\} \mid \lambda) &= 1.00 \cdot 0.70 \cdot 0.10 \cdot 0.6 \cdot 0.4 \cdot 0.1 = 0.00168 \\
P(O, Q = \{srr\} \mid \lambda) &= 1.00 \cdot 0.70 \cdot 0.10 \cdot 0.6 \cdot 0.6 \cdot 0.6 = \mathbf{0.01512} \\
P(O, Q = \{rss\} \mid \lambda) &= 0.00 \cdot \cdots \qquad = 0 \\
P(O, Q = \{rsr\} \mid \lambda) &= 0.00 \cdot \cdots \qquad = 0 \\
P(O, Q = \{rrs\} \mid \lambda) &= 0.00 \cdot \cdots \qquad = 0 \\
P(O, Q = \{rrr\} \mid \lambda) &= 0.00 \cdot \cdots \qquad = 0.
\end{aligned} \tag{2.15}$$

The prisoner can determine that most likely the two following days after entering the prison have been rainy.

Principally, HMMs can be applied for solving three basic problems (Rabiner 1990; Duda et al. 2001; Fink 2008):

1. Evaluation
   The first problem addresses the *production probability*, which can be calculated given an HMM defined by the parameters $\lambda = (A, B, \pi)$ and an observation sequence $O = \{O_1, O_2, \ldots, O_T\}$. The evaluation problem determines, how likely the HMM λ did generate the observed output sequence O, as in the previously described weather example. This "production probability" measures how well a model matches a given observation sequence. Given more than one model, this allows to choose the model that best matches the observations—and by that classify a sequential observation.
2. Decoding
   Again, an HMM and a concrete observation sequence are given. The decoding task is to find the correct internal state sequence $Q = \{q_1, q_2, \ldots, q_T\}$ that produced the observed output sequence, in other words, to uncover the internal process. The joint probability $P(O, Q \mid \lambda)$ of the observation and the state sequence given the model needs to be maximized.
   This task is often referred to as *decoding* of the model, as the production of observation symbols by the model can be compared to *coding* the internal state sequence into an output.
3. Learning
   The *learning* task aims to adapt the parameters A, B, π in a well-defined HMM with a number of n hidden states so that $P(O \mid \lambda)$ is maximized. The solution is a step-by-step improvement of the free parameters in the model in a way that it reliably describes the statistical properties of the appointed data. This is usually also denoted as *training* in ML.

Evaluation

A straightforward way to calculate the production probability is to find $P(O, Q \mid \lambda)$ for all possible state sequences Q and sum them up:

$$P(O \mid \lambda) = \sum_{Q} P(O \mid Q, \lambda) P(Q \mid \lambda) = \sum_{Q} P(O, Q \mid \lambda). \qquad (2.16)$$

This method is rather ineffective regarding computational complexity, since the number of possible state sequences increases exponentially with increasing sequence length. Therefore efficient algorithms are needed for practical applications. A frequently used one is the recursive *forward algorithm* (Duda et al. 2001) using the Markov property of the underlying chain. A *forward variable* α is defined as the probability of the given observation sequence up to time t, the state i given at time t and given the model λ:

$$\alpha_t(i) = P(o_1, o_2, \cdots, o_t, q_t = i \mid \lambda) = \begin{cases} \pi_i b_i(O_t) & \text{for } t = 1 \\ \left(\sum_j \alpha_t(j) a_{ji}\right) b_i(O_{t+1}) & \text{otherwise.} \end{cases} \qquad (2.17)$$

For a given observation sequence of length T, the forward variables for every possible end-state are calculated recursively by summing up the probabilities of being in state j before and transit to state i now, where observation o_t is emitted.

As Q_t is a set of mutually exclusive and exhaustive events, it can be easily shown that the sum of the forward variables for every of N states equals the production probability:

$$P(O \mid \lambda) = \sum_{i} P(O \mid q_T = i, \lambda) P(q_T = i) = \sum_{i} P(O, q_T = i \mid \lambda) = \sum_{i=1}^{N} \alpha_T(i) \qquad (2.18)$$

In Sect. 4.2.2 we will use this approach to model the user's emotional history in a dialog, which will allow us to ultimately estimate the user's current emotional state.

Decoding

Decoding has already been demonstrated in the weather example. Therefore, all possible state sequences were determined and every posterior probability of the given observation sequence was calculated. The state sequence Q^* with the maximum probability was choosen to picture the hidden weather conditions:

$$Q^* = \arg\max_{Q} P(O, Q \mid \lambda). \qquad (2.19)$$

Similar to the evaluation task, this brute force method is fairly ineffective. Exploiting the Markov property again allows the use of efficient algorithms. A common dynamic programming algorithm is the *Viterbi Algorithm* (Viterbi 1967; Fink 2008), which finds the optimum sequence with maximum posteriori likelihood. It is also frequently used in telecommunications.

Briefly described, it operates similar to the forward algorithm, with the difference that the summation step is replaced by maximization. Since the global maximum can only be determined at the end of the observation sequence, an additional variable for backtracking is defined, holding the dedicated optimal predecessor state.

Learning

HMM learning aims to identify the optimum model parameters based on a set of training samples. There is no method known for analytically obtaining these parameters in an optimal or most likely way (Pfister and Kaufmann 2008; Fink 2008). Instead, iterative techniques can be applied that find a local maximum of the production probability. Well-known techniques are the Forward-Backward algorithm (Rabiner and Jaung 1993; Pfister and Kaufmann 2008) and the Baum-Welch-Algorithm (Baum et al. 1970; Pfister 2008). A detailed description of HMMs can be found in Rabiner (1990).

2.1.2 Performance Metrics

Supervised learning aims to perfectly imitate and model statistical random processes. Due to the complexity of some tasks, the lack of training material and the inadequacy of some learning schemes, errors do occur. The various performance metrics used to assess the error-rate and the performance of classifiers as well as regression models are discussed in the following. That followed, we discuss the inter-rater agreement as measurement to evaluate the congruence of human annotators that manually assign labels to unlabeled data.

Metrics for Binary Classification

When evaluating the performance of a *binary* classifier (i.e. a classifier that may discriminate between two classes) with a test set of n samples, four numbers play an important role:

- True Positives (TP): the number of correctly classified samples from the positive class,
- False Positives (FP): the number of mistakenly as belonging to the positive class classified samples,
- True Negatives (TN): the number of correctly classified samples from the negative class and,

- False Negatives (FN): the number of incorrectly as belonging to the negative class classified samples.

The terms "positive" and "negative" originate from the field of information retrieval where "positive" stands for the relevant documents that a search algorithm should deliver whereas "negative" stands for irrelevant documents. As we will see later in Chap. 4, where we discuss the classification of angry speech samples versus non-angry speech samples, we will consider "anger" as positive class and "non-anger" as negative class.

It is easy to comprehend that the sum TP + FP + TN + FN represents the number of samples in the test set n. A standard evaluation metric is the accuracy, which is defined as:

$$\text{accuracy} = \frac{\text{TP} + \text{TN}}{\text{TP} + \text{TN} + \text{FP} + \text{FN}} \tag{2.20}$$

for evaluating binary classifiers.

The accuracy does not take into consideration a skew class distribution. For instance, consider an emotion recognition task where a classifier determines between two classes c_{angry} (the positive class) and $c_{\text{non-angry}}$ (the negative class). Let us further assume that 10 % of the samples belong to the class c_{angry} and the remaining 90 % to the class $c_{\text{non-angry}}$. It should be noted that an accuracy measurement often leads to misleading performance values as it follows the majority class to a greater extent than it follows other classes. For example, a classifier that recognizes non-angry samples well in this dataset, which is strongly skew towards non-angry samples, but fails in recognizing the few angry samples, would yield a high accuracy. The overall performance on the task, however, would be weak despite the high accuracy. For skew datasets, it is thus more appropriate to regard each class individually. Thereby, two performance measures are applied, the precision and the recall. The precision is defined as:

$$\text{precision} = \frac{\text{TP}}{\text{TP} + \text{FP}}. \tag{2.21}$$

Literally speaking, the precision expresses how many predictions that have been assigned to the positive class indeed belong to the positive class, i.e. it quantifies the percentage of the correct positive predictions.

Usually the precision is only defined for one class, namely the positive one. For some tasks it is relevant to further capture the precision of the negative class. It can be defined analogously:

$$\text{precision}_{\text{neg}} = \frac{\text{TN}}{\text{TN} + \text{FN}}. \tag{2.22}$$

The recall is defined as:

$$\text{recall} = \frac{\text{TP}}{\text{TP} + \text{FN}} \tag{2.23}$$

and describes the proportion of all positive samples in the set that have been classified as positive. For an anger recognition task this value represents the proportion of angry speech samples that could be identified by a classifier. This measure can be extended for likewise evaluating the negative class:

$$\text{recall}_{\text{neg}} = \frac{\text{TN}}{\text{TN} + \text{FP}}. \quad (2.24)$$

Based on precision and recall, the *F-measure* can be calculated as weighted harmonic mean of precision and recall:

$$F = 2 \cdot \frac{\text{precision} \cdot \text{recall}}{\text{precision} + \text{recall}}. \quad (2.25)$$

It describes the overall performance of a classifier for predicting the positive class. We further define the F_{neg}-measure for a performance assessment of the negative class:

$$F_{\text{neg}} = 2 \cdot \frac{\text{precision}_{\text{neg}} \cdot \text{recall}_{\text{neg}}}{\text{precision}_{\text{neg}} + \text{recall}_{\text{neg}}}. \quad (2.26)$$

Finally, the f_1 measurement is defined as the arithmetic mean of F-measures from all K classes:

$$f_1 = \frac{1}{K} \sum_{k=1}^{K} F_k. \quad (2.27)$$

Experiments described in the Chaps. 4, 5 and 6 are mainly assessed using class-wise precision, recall and F measures as well as overall f_1 scores.

Metrics for Evaluating Multi-Class and Numeric Predictions

When more than two classes are involved in the classification process, other performance metrics are required. The *Weighted Average Recall* (WAR), also denoted as accuracy, can be defined as:

$$\text{WAR/accuracy} = \frac{1}{S} \sum_{s=1}^{S} \text{match}(R_s, H_s), \quad (2.28)$$

where S is the number of test samples, match a function that returns 1 if the hypothesis H matches the reference class R and 0 otherwise. It should be noted that the accuracy for this multi-class evaluation metric is likewise influenced by the class distribution and often returns misleading performance values in case of skew datasets. Nonetheless, we will provide accuracy scores in some of our evaluations since it depicts a standard evaluation metric in many classification tasks. A more precise evaluation metric for skew multi-class sets is the *Unweighted Average Recall* (UAR)

$$\text{UAR} = \frac{1}{K}\sum_{k=1}^{K}\frac{\sum_{i\in\{i|R_i=k\}}\text{match}(R_i, H_i)}{\sum_{i\in\{i|R_i=k\}} 1}. \tag{2.29}$$

Obviously, the UAR does not take into account the distribution of samples in the data. If applying the UAR for assessing classifier performance, all classes are considered equally important. For UAC and WAR see also Higashinaka et al. (2010b).

In contrast to discriminative classifiers, which predict distinct classes on a non-ordered nominal scale, regressive classifiers predict continuous values on continuous or ordinal scales. Consequently an error is not present or absent, but comes in different sizes (Witten and Frank 2005). Hence, the distance of the hypothesis to the reference label is relevant, namely 5 is closer to 4 than to 1. Most commonly used in regressive evaluation tasks is the *Mean Squared Error* (MSE):

$$\text{MSE} = \frac{1}{n}\sum_{i=1}^{n}(R_i - H_i)^2. \tag{2.30}$$

It measures the mean deviation of all n hypothesis values (H) to the reference values (R).

The MSE weights large errors more heavily than small errors due to the nature of squaring the error. The also commonly used *Mean Absolute Error* (MAE) considers errors equally:

$$\text{MAE} = \frac{1}{n}\sum_{i=1}^{n}|R_i - H_i|. \tag{2.31}$$

Spearman's ρ is used to measure the correlation between two variables and is applied for ordinal scaled variables:

$$\rho = \frac{\sum_{i=1}^{L}(R_i - \overline{R})(H_i - \overline{H})}{\sqrt{\sum_{i=1}^{L}(R_i - \overline{R})^2(H_i - \overline{H})^2}}, \tag{2.32}$$

where R denotes the reference and H the hypothesis. A ρ value of 1 would denote perfect positive correlation, zero indicates that there is no tendency for R to either increase or decrease when H increases, and -1 would denote a perfect negatively correlated dependency between H and R.

Inter-Rater Reliability

In supervised learning it is required that the class labels of a data set are given. In case the labels are unknown, they have to be determined in a manual annotation process. For example, in a handwriting recognition task, letter labels have to be assigned manually to handwritten sample images of the different letters. For annotating an audio data set with gender labels, where the task is to choose between the labels "male" and "female", a single rater would generally suffice since gender is to a certain extent unambiguous. For example, when annotating emotions or the estimated age of the speaker contained in an audio data set, the annotation task gains complexity. A reliable label can thus only be obtained by several expert annotators.

To obtain a measurement for consistency of such impressions an inter-rater agreement measurement can be applied. Informally speaking, it is the ratio of the chance level corrected proportion of times that the labelers agree to the maximum proportion of times that the labelers could agree (Gwet 2010).

For measuring the inter-labeler agreement between two labelers, the Cohen's Kappa statistic (Cohen 1960) is used, for more labelers Fleiss' Kappa (Fleiss 1971) is employed. Cohen's Kappa κ_C is calculated as:

$$\kappa_C = \frac{p_A - p_R}{1 - p_R}, \tag{2.33}$$

where p_A is the agreement between the two labelers and p_R the expected agreement if both labelers would annotate on a random basis. The value of κ_C ranges from $-\infty$ (hardly any agreement) to 1 (total agreement). Fleiss's Kappa κ_F is calculated as:

$$\kappa_F = \frac{\bar{P}_0 - \bar{P}_d}{1 - \bar{P}_d}, \tag{2.34}$$

where $\bar{P}_d$ is the agreement by chance and $\bar{P}_0$ the average agreement between the labelers. So the factor $1 - \bar{P}_d$ gives the degree of agreement that is attainable above chance, and $\bar{P}_0 - \bar{P}_d$ gives the degree of agreement actually achieved above chance. Analog to κ_C, κ_F ranges from $-\infty$ (hardly any agreement) to 1 (total agreement). According to Landis and Koch (1977) the κ values of Cohen and Fleiss can be interpreted as follows:

- < 0: no agreement;
- $0.21 - 0.40$: fair agreement;
- $0.41 - 0.60$: moderate agreement;
- $0.61 - 0.80$: substantial agreement;
- $0.81 - 1.00$: almost perfect agreement.

It should be noted that an agreement on some tasks is easier to reach than on others: the agreement of rating the gender of speakers will yield higher rates between labelers than the agreement in an anger-/non-anger rating task. A low agreement on the different categories in the training- and test-data often implies a low classifier performance. This seems obvious since in cases where humans can only with difficulty agree on which category a sample belongs to, the classifier itself will likewise have trouble in differentiating between the blurred patterns.

2.2 Emotion Recognition

Reeves and Nass (1996) could show that users react emotionally when interacting with computers. If a computer makes a mistake, due to false user behavior or system failure, users tend to become annoyed and show their emotional feelings towards the machine. This also affects the opposite: when a computer pays compliments to the users, they tend to become delighted. As we have seen in the examples in Tables 1.1 and 1.2 (Sect. 1.2), emotions can also be observed in interactions with SDS. Critical situations like these could potentially be prevented if the dialog system would be able to detect that the users are becoming frustrated. This could either happen in an online manner (real time) in which the system permanently monitors the emotional state of the caller and escalates to a human operator once such a problematic situation is detected, or it can be done offline on logged and recorded dialogs to spot dialog steps where users frequently get angry. One can see how emotion detection, and specifically anger detection, could also be of great benefit when system developers try to find out flaws in the dialog design for the purpose of improving the overall system.

Estimating the user's emotional state is a typical classification problem and research has embraced a number of information sources to characterize and recognize emotional user behavior. Emotion recognition research ranges from gestures and facial expressions (Martin et al. 2006) or physiological measurements, such as electrocardiogram, skin temperature variation and electrodermal activity (Picard 2000; Bosma and André 2004; Kim et al. 2004), to speech (Batliner et al. 2000; Devillers et al. 2003; Schuller et al. 2009; Pittermann et al. 2009). For the employment of emotion recognition in SDS, not all techniques offer a sustainable way. For today's real-life applications, physiological measurements are unworkable as sensors would have to be attached to the user. Exploiting gestures and facial expressions would require visual information, which is commonly not available in SDS—and the least of all in telephone-based SDS. In this book we will thus concentrate on the recognition of emotions using speech and language information. However, we will further extend existing approaches by including interaction-patterns and contextual information.

This section gives an overview of related work in the field of speech-based emotion recognition with a main focus on the detection of emotions in real-life data. Common acoustic/prosodic and linguistic characteristics are presented that are frequently used in state of the art approaches.

2.2.1 Theories of Emotion and Categorization

Controversial results may be obtained when analyzing the taxonomy of the "emotional landscape". Scherer describes the attempts to definite emotions as "notorious problem" (Scherer 2005) . The first theories on emotions date back to the late nineteenth century. In "The Expression of the Emotions in Man and Animals", Charles

Darwin postulated that emotional expressions inherently belong to human beings by birth (Darwin 1872). They are automatic reactions on external events and phenomena, which played an important part in solving problems faced by the species. According to Darwin, emotions enhance its survival chance and therefore enhance the chance for being selected for reproduction. Hereby, human beings and to some degree animals share more or less the same emotions.

Darwin identifies and analyzes "low spirits" (anxiety, grief, dejection and despair) and "high spirits" (joy, love, tender feelings and devotion). Further he distinguishes and analyzes, among others, scorn, disdain, contempt and disgust as well as guilt, pride, helplessness, patience and affirmation. Another common emotion theory originates from David James, who stated in 1884 in his article "What is emotion?":

> "I should say first of all that the only emotions I propose expressly to consider here are those that have a *distinct bodily expression*. That there are feelings of pleasure and displeasure, of interest and excitement, bound up with mental operations, but having no obvious bodily expression [..]." (James 1884).

For both, Darwin and James, emotional expressions are first of all physiological sensations helping to survive, whereas James sees in emotions reactions or feelings to physical changes of the body. For James, emotions are results of these changes, not their cause.

According to Cornelius, the "cognitive perspective" can be considered as the dominant perspective (Cornelius 2000). He postulates that emotional expressions imply an appraisal, i.e. the positive or negative judgment of events in the environment (Arnold 1960). For the cognitivists, emotion and thought are inseparable. The appraisal triggers emotions and provides the link between particular characteristics of a person, the learning history, temperament, personality, physiological state and the characteristics of the inputs from the environment.

Finally, the "youngest, most diverse, and certainly most controversial of the four theoretical perspectives" (Cornelius 2000) is the social constructivist perspective, which postulates that

> "[..] emotions are cultural products that owe their meaning and coherence to learned social rules" (Cornelius 2000).

Here emotions, such as fear or anger, have a specific purpose in our social and moral understanding.

In contrast to other affective states such as moods, interpersonal stances, preferences/attitudes and affect dispositions, emotions are intense and of short duration (Scherer 2000).

In emotion research there is no gold-standard that predominates when it comes to categorize emotions of human subjects. No agreement exists on a general set of basic emotions (Scherer 2005), however, a number of categorizations exist that include such "basic", "primary" or "fundamental" emotions. According to Cornelius, these emotions are considered fundamental because "they represent survival-related patterns of responses to events in the world that have been selected for over the course of our evolutionary history." (Cornelius 2000). Plutchik (1980) introduced

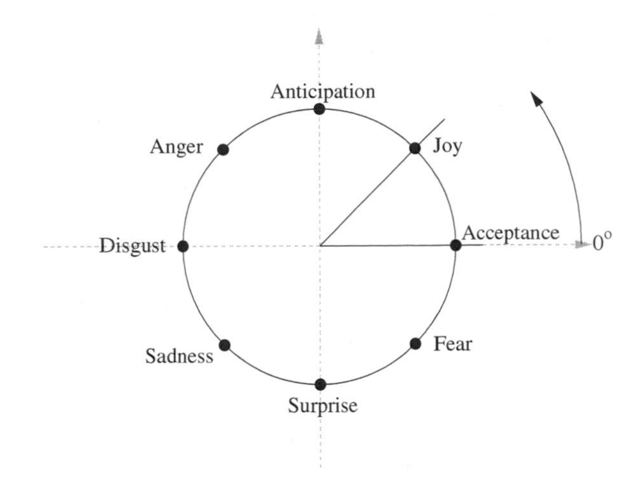

Fig. 2.6 Emotion wheel (Plutchik 1980; Pittermann et al. 2009)

the "emotion wheel", to categorize basic emotions (see also Cowie et al. (2001), Pittermann et al. (2009)). Plutchik arranges the emotions in a circle, see Fig. 2.6, where each basic opposite is positioned on the opposite side of the wheel.

Their characteristics are here reflected by angular measures where, e.g. 0° represents agreement, 90° anticipation, 180° disgust and 270° surprise. While Plutchik suggests the eight basic emotions—acceptance, joy, anticipation, anger, disgust, sadness, surprise, fear—Cornelius proposes the "Big Six": happiness, sadness, fear, disgust, anger, and surprise (Cornelius 1996). Other definitions of basic emotions sets can be found in Schlosberg (1954), Russell (1980), Ekman (1992), Nisimura et al. (2006).

When implementing emotion recognition in SDS with the ultimate aim to achieve adaptivity, the particular benefit of including a specific emotion class into the recognition process should be estimated in advance. For example, there is little point in recognizing disgust without defining a specific action that follows the new knowledge that the user is disgusted. Emotions not relevant for adaptivity should thus not be considered in the training process as with a larger number of classes, the risk of confusion likewise increases.

2.2.2 Emotional Speech

The implementation of emotion recognition techniques requires training corpora. Campbell et al. (2006) discriminate between two types of such data sets: *artificial corpora*, i.e. acted or induced data, and *real-life spontaneous corpora*, i.e. authentic speech methods. Pittermann et al. (2009) finds a more differentiated taxonomy. He characterizes emotion corpora according to the following criteria:

- Control of the speech characteristic through the observer (i.e. the person collecting the speech material) or no control.
- In vitro method (laboratory corpus) or in vivo method.

- Professional actors or non-professional speakers.
- Utterances linguistically and phonetically predefined or undefined.
- Utterances with emotional content or semantically neutral.

According to Schuller (2006), the challenges in recognizing emotions depend on different aspects: acted versus spontaneous speech, realistic emotions, noise and microphone conditions, and speaker independence. We have to add another important aspect, namely the average length of the speech samples. Samples of longer duration are easier to classify than very short samples, since in that case, less information can be derived from the audio signal.

Studies on emotion recognition started with acted corpora containing so called full-blown emotions, see e.g. Enberg and Hansen (1996), Douglas-Cowie et al. (2000), Burkhardt et al. (2005b). The Danish Emotional Speech (DES) database Enberg and Hansen (1996) consists of utterances from two male and two female actors who were asked to convey a number of emotions, which are surprise, happiness, sadness and anger as well as the neutral state. Twenty listeners had to identify the emotions presented in the recordings and identified 67.3 % correctly. Douglas-Cowie recorded the "Belfast"-Corpus, that consists of 30 acted audio-visual studio recording clips plus 209 clips from a variety of TV shows, such as chat shows and religious programs. The Database of German Emotional Speech (Burkhardt et al. 2005), also denoted as "Berlin Corpus", consists of 800 spoken utterances from five female and five male prevailingly professional actors that were recorded while reading sentences in an emotional way. The recorded emotion set comes close to the "Big Six" (Cornelius 2000) and consists of anger, sadness, joy, disgust, fear, boredom as well as the neutral state. The 48 kHz recordings have been down sampled to 16 kHz which depicts a considerable higher quality than e.g. μ-law compressed telephone data.

Many previous studies dealing with the recognition of emotions in speech use acted corpora for their experiments. While they provide insight into the acoustic differences of emotional speech, they can hardly be compared to studies using real-life corpora. When classifying speech captured under real-life conditions a number of differences are visible in comparison to acted corpora:

- Lower sampling rates and compressed audio, when speech is transferred over telephone-channels. This especially hampers a reliable extraction of the fundamental frequency and other acoustic description patterns.
- Varying distances to the capturing device, e.g. built-in microphone in a moving robot (Batliner et al. 2004), various speaker positions when interacting with interactive kiosk systems and telephone systems (e.g. hands-free interaction).
- Background noise from outdoor environments such as subway, bubble, street noise, but likewise noise from domestic settings, such as barking dogs, crying children, cross-talk, etc.
- Strong unbalanced occurrence of specific emotion classes and a usually high rate of neutral speech.

Classification results obtained with acted corpora can thus not be compared to results from realistic data. Schuller et al. (2007) found clear difficulties when classifying speaker-independent, spontaneous speech under noisy conditions in comparison to acted data in studio-quality.

The purpose of this book is to use emotion recognition for detecting critical dialog scenarios. The estimation of the user's emotional state other than annoyance and anger, such as happiness, sadness, fear, disgust and surprise would not help to accomplish this. The data employed for this work originates from real-life telephone applications, where users interact with a computer system over telephone where such emotional states do most likely not occur, or at least if they occur, they play only a very subordinate role. According to the characteristics previously described this speech data belongs to the group that is most difficult to classify: a large proportion of neutral speech, short speech samples recorded in vivo, i.e. in real-life from non-professional speakers under noisy conditions. With an increasing number of possible emotional states to be detected, the performance of the classifier drops, as the probability for confusions with the other emotion classes increases likewise. As a consequence, emotion detection in this work will concentrate on recognizing *anger* versus *non-anger* and will thus summarize other emotional states with no obvious relevance for the task under a joint class label.

The following list of emotional speech databases provides an overview of existing real-life speech corpora of narrow-band telephone quality. While this list raises no claim to completeness, it provides an overview of the existing corpora employed in emotion research specialized on the modeling of *negative* speech for *real-life* applications, i.e. the presented corpora are, similarly to this work, focused on the recognition of anger. For a comprehensive list of speech-databases that includes acted speech and recordings from Wizard-of-Oz[1] scenarios, refer to Pittermann et al. (2009).

CEMO Corpus Human-human dialogs of emergency calls, narrow-band telephone quality (Devillers and Vidrascu 2006)

Language: French
Emotions: Fear, Anger, Sadness, Hurt, Surprise, Relief, Other Positive, Neutral
Annotation: 2 annotators, combination using soft emotion vector
Size: 688 dialogs, 20 h.

AT&T HMIHY 0300 (***How May I Help You?***) Human-machine dialogs of AT&T customers interacting with the system, narrow-band telephone quality. Other studies dealing with this corpus: Shafran et al. (2003), Shafran (2005); other labels: age, dialect, gender

Language: English
Emotions: PosNeu (positive, neutral, cheerful, amused or serious), SWAngry (somewhat angry), VAngry (very angry, as with raised voices), SWFrstd (some-

[1] In a Wizard-of-Oz experiment the user assumes to interact with a more or less intelligent computer system. In reality, a hidden person simulates the intelligence of the system without the probands knowledge.

what frustrated), VFrstd (very frustrated), OtherSWNeg (otherwise category of lower intensity) and OtherVNeg (otherwise category of higher intensity)
Annotation: 1 annotator, combination using soft emotion vector 2 annotators labeled a subset of 200 dialogs for consistency PosNeu, SWAngry, SWFrstd, OtherSWNeg, VAngry, VFrstd: $\kappa = 0.32$ PosNeu, {SWAngry, SWFrstd, OtherSWNeg}, VAngry, VFrstd: $\kappa = 0.34$ PosNeu, {SWAngry, SWFrstd, OtherSWNeg, VAngry, VFrstd} $\kappa = 0.42$
Size: 5,147 utterances, 1,854 calls, 15 words per utterance on the average.

AT&T HMIHY 0300 (***How May I Help You?***) Human-machine dialogs, narrow-band telephone quality (8 kHz, 8 bit μ-law) compressed (Liscombe et al. 2005)

Language: English
Emotions: positive/neutral, somewhat frustrated, very frustrated, somewhat angry, very angry, somewhat other negative, very other negative
Annotation: 3 annotators; $\kappa = 0.32$; $\kappa = 0.42$ when collapsed to positive/neutral versus other
Size: 5,690 dialogs, 20,013 utterances.

SpeechWorks call-center corpus human-machine dialogs, narrow-band telephone quality (8 kHz, 8 bit μ-law) compressed (Lee et al. 2001, 2002; Lee and Narayanan 2005)

Language: English
Emotions: negative, non-negative
Annotation: 4 annotators; $\kappa = 0.465$
Size: 1,187 dialogs, 7,200 utterances, 6 utterances per call on the average; 1,367 utterances used in experiments.

T-Mobile call-center corpus Human-machine dialogs; call router application (Burkhardt et al. 2008)

Language: German
Emotions: 1: not angry, 2: not sure, 3: slightly angry, 4: clear anger, 5: clear rage
Annotation: 3 annotators; $\kappa = 0.52$; combination using mean values of all raters and thresholds
Size: 4,683 dialogs, 26,970 utterances, 5.8 utterances per call on the average; 50 % of the calls contain 3 utterances and less.

Most mentioned corpora contain merely audio recordings from user prompts with manually transcribed utterances. Immensely important interaction features giving evidence for example over user interruptions, system actions and temporal information are not included. Furthermore, the corpora in this field are all from commercial systems and may thus not be shared for research. As a result of this, the collection and annotation of new dialog corpora is unavoidable and has been conducted in the last years during our work.

2.2.3 Emotional Labeling

When speakers are not acting, i.e. in cases where there is no professional performance, the class label for a speech sample is unknown and requires manual annotation. There are several methods of annotating (also "labeling" or "rating") the emotions in a speech corpus. Stibbard (2001) discriminates between free choice of the labels, i.e. the labels for the utterances are freely chosen by the annotators, and forced choice, i.e. a predefined set of labels is given, among which the annotators have to choose from. The first approach can be considered as not expedient as it may result in a very large number of classes preventing that certain emotions are differentiated and recognized reliably. For large data-sets, this further imposes a time-consuming task to the rater. A post-processing step would be required that merges the large number of the rater's labels to distinct classes. When applying the latter procedure these emotion classes have to meet the requirements for the specific emotion recognition task and require a thorough definition. Here, it also should be kept in mind that similar labels make it difficult for both the rater and the classifier to differentiate and recognize emotions reliably. Furthermore, the labels can be of one-dimensional or multidimensional, continuous character, such as the degree of joy or the intensity of activation, valence and arousal.

It further has to be decided how the corpus is labeled. There are three approaches Pittermann et al. (2009):

- *Manual labeling* constitutes the most time-consuming but also the most reliable of these methods. Since one or more raters listen to every utterance and decide into which emotion class it belongs to, this method provides a realistic classification of the utterance from a human point of view.
- *Semi-automatic labeling* combines manual labeling with training a labeling classifier (cf. Fig. reffig:bootstrap). The procedure is as follows: first, a subset of the data is labeled manually, this subset is then employed for training a classifier, which uses the computed model to classify and annotate another, unlabeled subset of the data. This subset is then reviewed by a human expert for accuracy and correctness and adjusted if necessary. Next, the classifier is provided with the new labeled set of data and computes a new, improved model, which is used to classify another subset. This process is iterated until all the data is annotated (Pittermann et al. 2009; Abdennadher et al. 2007). Semi-automatic labeling can particularly be of valuable help for visual labeling tasks, e.g. for annotating letters in an Optical Character Recognition (OCR) task. For audio annotation, this procedure can be considered as less appropriate as it implies that, despite the automatic part, every utterance has to be manually examined and listened to.
- *Automatic labeling* relies heavily on the precondition that detailed information about the properties of the labels already exists. Hence, for automatic labeling of emotions, a deep insight of the acoustic properties of the emotions would be necessary. However, not having this insight is one of the presumptions for building classifiers. Otherwise there would be no need to apply a classifier on labeled data

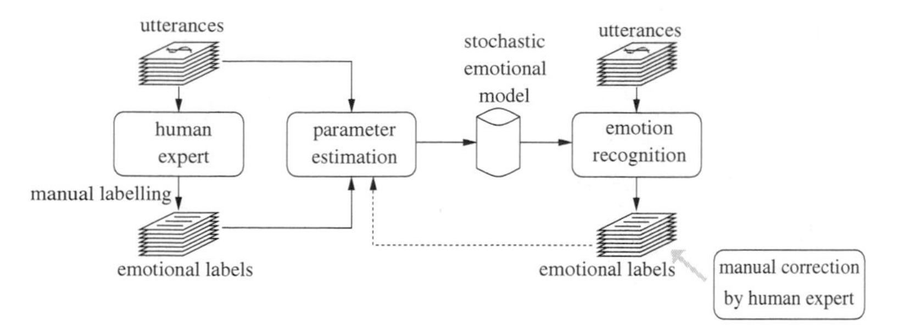

Fig. 2.7 Bootstrapping an emotion corpus (Pittermann et al. 2009)

to derive a model, which classifies the data. Therefore, automatic labeling is a step which can be applied when good classification models are already available.

As the labeling varies with the annotator it is suggested that the annotator is especially trained in labeling (Campbell et al. 2006). On the other side, it is also desired that the annotator labels in a normal intuitive way, which might be difficult to achieve if trained thoroughly before. As a solution, multiple labelers come to hand, annotating the data separately. Since no agreed-upon common opinion exists on how a specific emotion 'sounds' like it has become best practice to take into account the opinion of several raters. The aim is then to draw a clear line in between the emotion classes. Samples where no agreement on the emotion class is reached have to be sorted out. To obtain a measurement for consistency of such impressions the inter-rater agreement measurement can be applied. It is the ratio of the chance level corrected proportion of times that the labelers agree to the maximum proportion of times that the labelers could agree. A formal definition of the inter-rater reliability has been provided in Sect. 2.1.2. It should be noted that an agreement on some tasks is more easily to reach than on others: the agreement of rating the gender of speakers will yield higher rates between labelers than the agreement in an anger-/non-anger rating task. We will describe these effects in Sect. 3.2.2, where we compare κ-values of gender annotations with those from emotion annotations. It can be assumed that a low agreement on the different emotion categories in the training- and test-data implies low classifier accuracy. This seems natural since in cases where humans can only with difficulty agree on which emotion category a speech sample belongs to, the classifier itself will likewise have trouble in differentiating between the blurred patterns.

This and the frequently lower speech quality can be considered as main reasons why studies dealing with non-acted speech report much lower performance values as studies working on acted speech recorded under controlled conditions.

In the literature, different annotation schemes have been reported for annotating negative emotions.

- Ang et al. (2002) labeled utterances from the DARPA Project as frustrated, annoyed, tired, amused, neutral and other. The data, consisting of about 21,000 utterances, was provided by users who tried to make travel arrangements with an automated agent. Although the callers were not actors, they never intended to make real arrangements, so the occurrence of emotions may be somehow different to real-life data. The first labeling resulted in a Kappa value of 0.47, which the authors deemed too low for their purposes and conducted a second labeling by two of their most experienced labelers.

- A similar approach of classification of negative and non-negative emotions is also discussed by Lee et al. (2002). The authors report a 65 % agreement after annotation of the speech corpus. The rating process is based on two raters annotating a binary label of *negative* and *non-negative* emotions. It should be noted that the percentual agreement does not consider an agreement by chance and by that achieves higher scores.
- Liscombe et al. (2005) classified negative against neutral emotions. The corpus HMIHY300 (How May I Help You) is provided by AT&T's "How May I Help You" natural language human-computer spoken dialog system. 5,147 utterances were annotated with the following labels: positive/neutral, somewhat frustrated, very frustrated, somewhat angry, very angry, somewhat other negative, very other negative. The Cohen's Kappa statistic was 0.32 for all classes and 0.42 when the classes were collapsed to positive/neutral versus other.
- In a recent approach (Burkhardt et al. 2009) annotated 21 h recordings from a German voice portal. The 26,970 utterances were labeled by three raters with one of the following labels: 1: not angry, 2: not sure, 3: slightly angry, 4: clear anger, 5: clear rage or marked as "non applicable" if it is not suitable for any class (i.e. baby crying, TV in the background, etc.). These six labels were then collapsed into the labels "not angry", unsure", "angry" and "garbage" by calculating the mean label values of the three labelers ("non applicable" counts as 0). The authors report a Kappa value of 0.52 for all three labelers.

The challenge in creating an annotation scheme lies in keeping the balance between a detailed representation of a concept in adequate classes and the ease of the labeling process. More classes mean more choice for the rater, which slows down the annotation, whereas too few classes may mean a loss of information as it may force the rater to suboptimal decisions. As our interest lies in the detection of annoyance and anger, we define the classes SLIGHTANGER and STRONGANGER leaving room for different degrees of anger while keeping the choice for the rater limited at the same time. Furthermore, all other emotions are summarized as NONANGER. Non-speech events and other incomprehensive speech is summarized as GARBAGE. The proposed annotation scheme uses more classes than Lee et al. (2002), but fewer classes than (Burkhardt et al. 2009). Another important aspect is the choice of the number of raters. The more individual opinions make up the finally assigned class label, the more objective the label and the higher the quality of the resulting statistical model after training. A reasonable compromise between costs and quality is in our point of view the choice of three raters. Details are provided in Sect. 3.2.2.

2.2.4 Paralinguistic and Linguistic Features for Emotion Recognition

Paralinguistic description patterns (which are basically of prosodic and acoustic nature) and linguistic information can be applied when recognizing emotions from speech. In the following, common paralinguistic and linguistic features are presented.

Paralinguistic Recognition

Speech-based emotion recognition follows a similar procedure as speech recognition. The recorded audio signal has to be digitalized, e.g. in form of an uncompressed data stream (Eppinger and Herter 1993). Afterwards, signal analysis extracts acoustic parameters, such as cepstral coefficients, loudness, intensity, formants, etc. This is commonly accomplished by analyzing sub-parts or frames of the signal. For each frame, often representing 10 ms of speech, these acoustic description patterns are determined. In the next step, the data is reduced and subject to feature extraction, which results in a series of feature vectors, each representing a frame of the signal or a single static feature vector, representing the entire utterance. It should be noted that the size of this vector does not depend on the length of the utterance and is thus called "static". Hereby the dynamics of the speech signal, i.e. the changing acoustic contours, have to be modeled and stored as separate features in the static vector to ensure that this temporal information is not lost. Whether a static feature vector or a series of vectors are required depends on the classification algorithm. HMMs demand sequential feature vectors while other pattern recognition approaches such as KNN, Artificial Neural Networks (ANN) and SVMs, can only handle a single vector. Following the feature extraction, a preprocessing step can be applied to normalize the feature vector and to reduce redundancies. The normalized vector, that represents the characteristics of a speech sample, is then used for classification. Figure 2.8 gives an overview of this process.

Prosody

The acoustic features that are predominantly employed for emotion recognition are the prosodic features (Panctic and Rothkrantz 2003). Prosody is a characteristic of speech which comprises intonation, rate and change in loudness of a spoken utterance. Absence of prosody in spoken language results in alienated and desultory speech, which is hard to understand. Therefore it is a very important indicator for emotions. Prosody is mainly influenced by the fundamental frequency, syllable length, loudness, pitch, and the formants (Epinger and Herter 1993). In general prosodic features summarize properties of melody and rhythm and are to some degree uninfluenced by specific language properties (Schuller 2006). Prosodic features consist of intensity

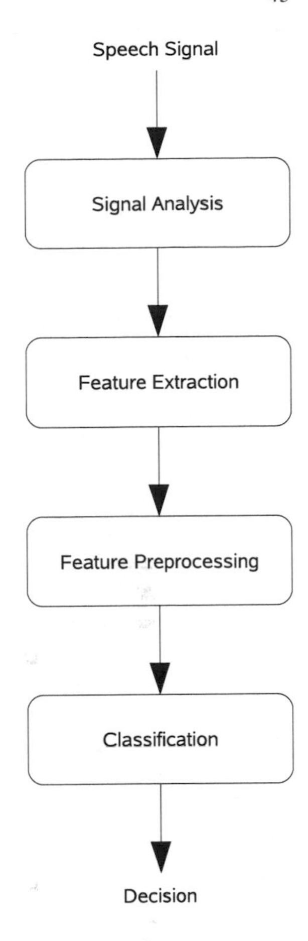

Fig. 2.8 Diagram of a pattern recognition system in speech processing (Eppinger and Herter 1993)

(the perceptual loudness of a sound), intonation (the pitch of a sound) and durational features, such as the ratio of voiced to unvoiced sounds. The *pitch* of a voice is "that attribute of auditory sensation in which sounds may be ordered on a musical scale" (American Standards Association). While frequency is an objective physical measure, pitch is a subjective perceptual measure. Strongly related to frequency, but not in one-to-one ratio the pitch is also influenced by intensity and waveform of the audio signal (Hall 1998). *Sound intensity* or *acoustic intensity* I of a sound wave is defined as "the average rate of flow of energy through a unit area normal to the direction of propagation". It is described by:

$$I = \frac{1}{t} \int_{0}^{t} p(t)u(t)\mathrm{d}t, \tag{2.35}$$

where $p(t)$ is the acoustic pressure and $u(t)$ the particle velocity (Kinsler et al. 1982).

The *sound power* or *acoustic power* Π is defined as:

$$\Pi = \frac{1}{t_2 - t_1} \int_{t_1}^{t_2} x^2(t)\mathrm{d}t \tag{2.36}$$

where t_1 and t_2 state the time frame in which the power is to be calculated and $x(t)$ is the amplitude of the sound. Sound power is neither room dependent nor distance dependent. Sound power belongs strictly to the sound source and is measured in Watts. Generally speaking, the higher the sound power, the louder the sound (Kinsler et al. 1982).

The alternation of voiced and unvoiced parts in the speech indicates the *speech rate*.

Voice Quality

Further acoustic features can be attributed to the voice quality, also denoted as "timbre". A speech signal normally consists of a harmonic mixture of frequencies. The *fundamental frequency* is the lowest, natural frequency of a speech signal. The other involved frequencies, which are multiples of the fundamental frequencies are called *overtones, harmonics or partials* (Kinsler et al. 1982). Where those harmonics supplement each other, locale maxima in the frequency are found, which are the *formants*. In Fig. 2.9 these maxima can easily be seen where the black color is densest. These bars represent the formants.

Harmonicity represents the degree of acoustic periodicity, also called Harmonics-to-Noise Ratio (HNR) and is used as a measure for signal-to-noise-ratio and voice quality (Boersma and Weenink 2009). Another two attributes describing the quality of the voice are *shimmer* and *jitter*, whereas jitter considers the variation of the fundamental frequency and shimmer the variations of the peak amplitudes (Pittermann et al. 2009).

Mel-Frequency Cepstral Coefficients

Mel-frequency cepstral coefficients (MFCC) (Rabiner and Juang 1993) are mainly employed for speech recognition but are also used for emotion recognition, see e.g. Schuller (2006), Pittermann et al. (2009), Burkhardt et al. (2009), Schuller et al. (2010). They depict a compressed form of the frequency spectrum. Since the human ear perceives frequencies non-linearly across the spectrum an empiric scale, the *mel scale*, was defined to approach the subjective perceptions of humans when listening to equidistant pitches. It is defined as:

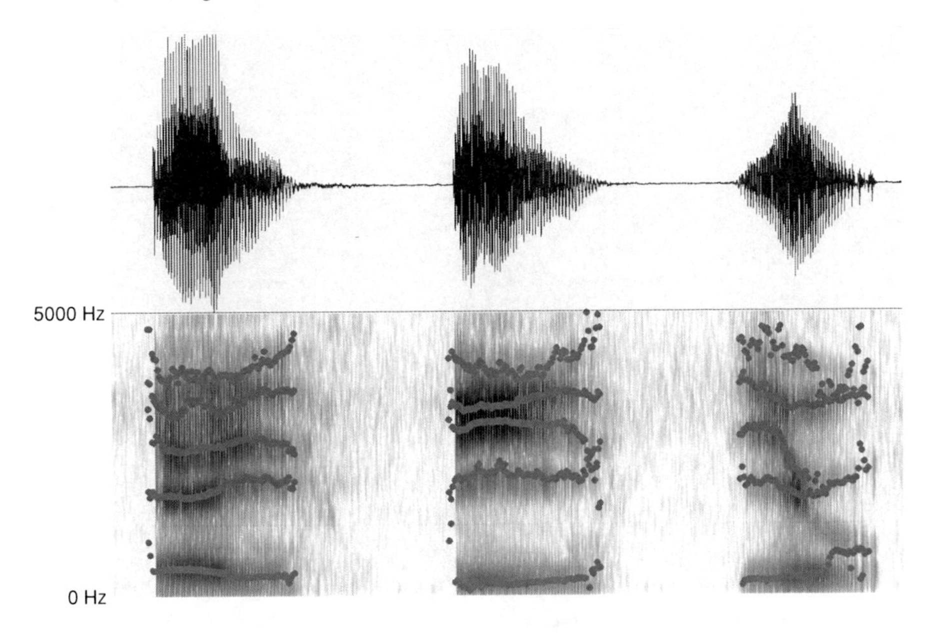

Fig. 2.9 A spectogram of the American English vocals a, e and u, showing the firat five formats

$$m = 2595\text{mel} \cdot \log_{10}\left(1 + \frac{f}{700\,\text{Hz}}\right), \tag{2.37}$$

where f is the frequency in Hertz.

This scale is used to calculate the mel-frequency spectrum by taking the Fourier transform of the signal and map the powers of the obtained spectrum onto the Mel scale, using triangular filters as shown in Fig. refmelfilterbank. From the mel-frequency spectrum, the mel-frequency cepstrum is received by taking the log of the powers at each of the mel frequencies followed by an application of a discrete cosine transform of the mel log powers. The MFCC are the amplitudes of the resulting spectrum Pittermann et al. (2009).

Linguistic Features

The emotional state of a speaker is prevailingly determined based on acoustic patterns. For example, the acted corpus German Database of Emotional Speech (Burkhardt et al. 2005b) has intentionally been designed in a way that the lexical content of the corpus utterances does not allow a deduction to an emotional state. The sentence "Der Lappen liegt auf dem Eisschrank" ("The cloth is lying on the fridge") or "An den Wochenenden bin ich jetzt immer nach Hause gefahren und habe Agnes besucht." ("At the weekends I used to go home and visit Agnes.") can be considered as completely unemotional from a linguistic point of view. Unlike this artificial

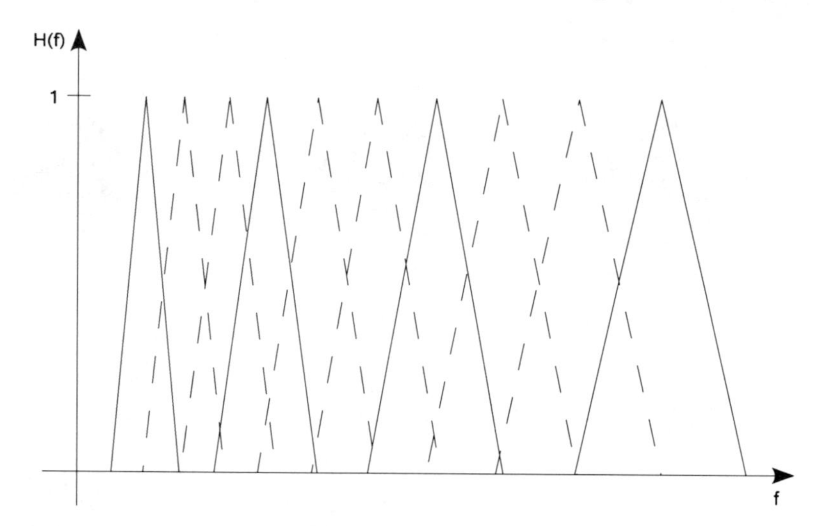

Fig. 2.10 Mel-scale filterbank (Pittermann et al. 2009)

Table 2.1 Emotional dictionary of linguistic emotion recognition (Pitterman and Schmitt 2008; Pittermann et al. 2009)

Keyword	Emotion	Valence
Acceptable	Happiness	++
Awful	Disgust	–
Boredom	Boredom	0
Cool	Happiness	++
Funny	Happiness	++
Irritated	Fear	–
Ok	Neutral	0
Ridiculous	Anger	–
Sorry	Sadness	–
Wow	Happiness	++
Yawn	Boredom	0
Yuk	Disgust	–

The first column contains the keywords, the second column contains an emotional state and the third column indicates the valence (positive/neutral/negative)

setting, certain words in non-acted, spontaneous speech co-occur more likely with a specific emotion than others, which is exploited in linguistic emotion recognition. Pittermann and Schmitt (2008) and Pittermann et al. (2009) used a hand-crafted emotional dictionary, see Table 2.1.

This list of emotional words is employed to assign an emotion to a spoken utterance based on linguistic content after parsing the sentence. While the proposed list might appear somewhat arbitrary there exist more elaborated approaches relying on data-driven techniques, e.g. the recognition of emotions based on unigrams. This is accomplished as follows.

The most likely emotional class $\hat{\varepsilon}$ of an utterance $\mathscr{U}$ consisting of the words w_i can be evaluated based on $P(\varepsilon_i|w_j)$, which is the a-posteriori probability of word w_j co-occurring with emotion ε_i and is obtained from a training corpus (Schuller 2006):

$$\hat{\varepsilon} = \arg\max_{\varepsilon_i} \prod_{w_j \in \mathscr{U} \wedge w_j \in \mathscr{V}} P(\varepsilon_i|w_j). \tag{2.38}$$

The most likely emotion class is the one that maximizes this equation. Unigrams are also employed in a similar form in Devillers and Vidrascu (2006).

Schuller (2006) explores further statistical methods using the Bag-of-Words representation, which is widely used for classifying documents. Hereby, all words in a training corpus represent the vocabulary $\mathscr{V}$ spanning a vector space of n dimensions, where n is the vocabulary size. It should be noted that the word-order in such vector representations gets lost. However, this is considered insignificant for both, document classification and linguistic emotion recognition.

Term Frequency

An utterance can now be represented as a feature vector of size n, where each feature component i in the vector represents the number of a each specific word in an utterance, normalized on the size of the utterance length $l = |\mathscr{U}|$:

$$x_{\mathrm{TF},i} = \frac{\mathrm{TF}(w_i, \mathscr{U})}{l}. \tag{2.39}$$

For example from a training corpus $\mathscr{S} = \{\mathscr{U}_1 =$"I want to speak to an operator", $\mathscr{U}_2 =$"Give me a representative"$\}$ we derive the vocabulary $\mathscr{V} = \{$a, an, give, i, me, operator, representative, speak, to, want$\}$ of size $n = 10$.

The utterance "I want to speak to an operator" would be coded as:

$$\begin{pmatrix} \frac{0}{7} \\ \frac{1}{7} \\ \frac{0}{7} \\ \frac{1}{7} \\ \frac{0}{7} \\ \frac{1}{7} \\ \frac{0}{7} \\ \frac{1}{7} \\ \frac{2}{7} \\ \frac{1}{7} \end{pmatrix} \begin{pmatrix} \text{a} \\ \text{an} \\ \text{give} \\ \text{i} \\ \text{me} \\ \text{operator} \\ \text{representative} \\ \text{speak} \\ \text{to} \\ \text{want} \end{pmatrix}. \tag{2.40}$$

It is easy to comprehend that such vectors can quickly gain dimensions of several thousand, depending on the size of the vocabulary. A large proportion of the vector components thus remain zero.

Term Frequency Inverse Document Frequency

Further, the document frequency (DF) determines in how many documents, here utterances, the specific word is observed. An inverse document frequency (IDF) can be determined:

$$\mathrm{IDF}(w_i) = \frac{L}{\mathrm{DF}(w_i)} \tag{2.41}$$

where L determines the number of utterances in the training set. The feature components of the vector when applying the DF are defined as:

$$x_{\mathrm{TF.IDF},i} = \mathrm{TF}(w_i, \mathcal{U}) \cdot \mathrm{IDF}(w_i) \tag{2.42}$$

Bag-of-Word and other vector space models have not been sufficiently taken into account when addressing linguistic emotion recognition. Particularly emotions occurring in real-life SDS in combination with such data-driven models require further consideration. More experiments are required to assess the validity and performance of statistical approaches to linguistic emotion recognition. This issue will be addressed in Sect. 4.1.2.

2.2.5 Related Work in Speech-Based Emotion Recognition

A considerable number of studies deal with the speech-based recognition of emotions in a wider sense. Studies related to the TU Berlin EMO-DB Burkhardt et al. (2005b) corpus based on speech produced by German speaking professional actors yield considerable performances when classifying emotions. The lexical content is limited to ten pre-selected sentences all of which are conditioned to be interpretable in six different emotions and neutral speech. The recordings have wideband quality. Experiments on a subset, exceeding human recognition rates of 80 % and human naturalness votes of 60 %, resulted in 92 % accuracy when Schuller classified for emotions and neutral speech (Schuller 2006). The recognition of angry utterances in this corpus yields an accuracy of 97 % in a seven-classed recognition test performed by humans (Burkhardt et al. 2005b). Pittermann et al. (2009) obtain 75.8 % when classifying six emotions using a combined speech-emotion recognizer.

Mostly comprising recordings from read sentences but also including free text passages a further anger recognition experiment was carried out on the DES database (Enberg and Hansen 1996). The accuracy for human anger recognition resulted in

75 % out of five classes. All recordings are of wide band quality as well. Classifying automatically Schuller obtains 81 % accuracy (Schuller 2006).

The ever increasing deployment of IVR applications opens possibilities to conduct studies based on real-life data. Early studies faced the problem that such field data was hardly available. Instead, the topic of anger detection was approached merely from an acoustic point of view with acted corpora or Wizard-of-Oz-based corpora containing high quality speech. The corpus coverage has grown from acted corpora with few speakers in high quality recorded in artificial surroundings to corpora containing several hundred speakers in narrow-band quality including interaction logs and ASR transcripts that constitute additional information sources.

One of the first studies addressing emotion recognition with special application to anger in telephone-based SDS was presented by Petrushin (1999) in 1999. Due to the lack of "real" data, the corpus was collected with 18 non-professional actors that were asked to record voice messages of 15–90 s in 22 kHz, which were split afterwards into chunks of 1–3 s. Raters were asked to label the chunks with "agitation" (including anger) and "calm". The recognition is performed exclusively on acoustic cues. Studies on real-life corpora came up in the early 2000s. Lee et al. (2001) employed data from a deployed IVR system, but complains about data sparsity. The dataset comprises a balanced set of only 142 short utterances with "negative" and "non-negative" emotions. In later studies (Lee et al. 2002; Lee and Narayanan 2005b) the employed set contained 1,197 utterances, however, strongly balanced towards non-angry emotions. Additionally to acoustic features, Lee et al. make use of linguistic cues that are extracted from ASR transcripts and propose the use of Emotional Salience as linguistic indicator.

Yacoub et al. (2003) state that real-life corpora are hard to obtain and employ an acted corpus containing eight speakers at 22 kHz quality. The utterance length is artificially shortened to simulate conditions typical for IVR systems. In Liscombe et al. (2005) the lexical and prosodic features were additionally enriched with dialog act features leading to an increase in accuracy of 2.3 %. Lee and Narayanan (2005) as well as Batliner et al. (2000) used real-life narrow-band IVR speech data from call centers. Both applied binary classification, i.e. Batliner discriminates angry from neutral speech, Lee and Narayanan classify for negative versus non-negative utterances. Given a two class task it is even more important to know the prior probability of class distribution. Batliner reaches an overall accuracy of 69 % using Linear Discriminative Classification (LDC). Unfortunately no class distribution or inter-labeler agreement for his corpus is given. Lee and Narayanan reached a gender dependent accuracy of 81 % for female and 82 % for male speakers. He measured inter labeler agreement with 0.45 for male and 0.47 for female speakers, which can be interpreted as moderate agreement. For both gender classes, constant voting for the non-negative class would mean to achieve roughly 75 % accuracy already and—without any classification—outperforms the results obtained by Batliner.

While the presented approaches aim to recognize emotions in single spoken utterances using acoustic, prosodic and lexical information they only have a limited benefit on measuring and evaluating the quality of a "conversation". In the following, we address related studies that aim to model system quality using interaction patterns.

2.3 Approaches for Estimating System Quality, User Satisfaction and Task Success

The dialog examples in Tables 1.1 and 1.2 (Sect 1.2) have shown that poorly performing interactions may exhibit other characteristics than the mere occurrence of emotional eruptions by the user. These properties may for example be continually recurring ASR errors or unexpected system prompts. Consequently, emotions are only one—but maybe not even the central symptom and aspect—that indicate fluent or non-fluent dialogs. Interaction patterns may therefore be used to estimate properties adhering to the interaction. In the following, we present related work in that field. We initially introduce approaches that use patterns obtained from a larger number of completed dialogs to estimate the quality of a dialog system. In contrast to these *offline* approaches, we also present—as what we call—*online* approaches that model interaction on the more fine-grained exchange level, aiming to predict user satisfaction and task success during the interaction.

2.3.1 Offline Estimation of Quality on Dialog- and System-Level

In software engineering, the evaluation phase has eventually become an important part of the development process which also affects the development of Spoken Dialog Systems (Dybjaer et al. 2007). While there exist structural recommendations on how to design a dialog system to meet user expectations (Bernsen and Dybkjaer 1997), the assessment of these systems is still often carried out intuitively (Möller 2005). Hone and Graham (2000) proposed such a structured approach consisting of a survey that is widely used to assess user satisfaction in SDS. This "Subjective Assessment of Speech System Interfaces" (SASSI) questionnaire has been applied in a number of user studies for evaluating SDS, e.g. Strauss and Minker (2010), Weiss et al. (2008), Gödde et al. (2008), and aims to capture user satisfaction as indicator for system quality. However, conducting user experiments implicates a large and costly effort for SDS developers. The aim of the research community in the recent past was thus to limit the need for such user tests and to find ways that allow an automatic estimation of system performance using data-driven techniques. The underlying idea is to statistically model system quality by training classification and regression models with dialog data serving as input variables, whereas the target variable is the actual user satisfaction originating from "real" users or expert annotators.

2.3.1.1 PARADISE

Walker et al. (1997, 1998, 2000) proposed the first and most prominent framework for automatically evaluating SDS, the "PARAdigm for DIalogue System Evaluation".

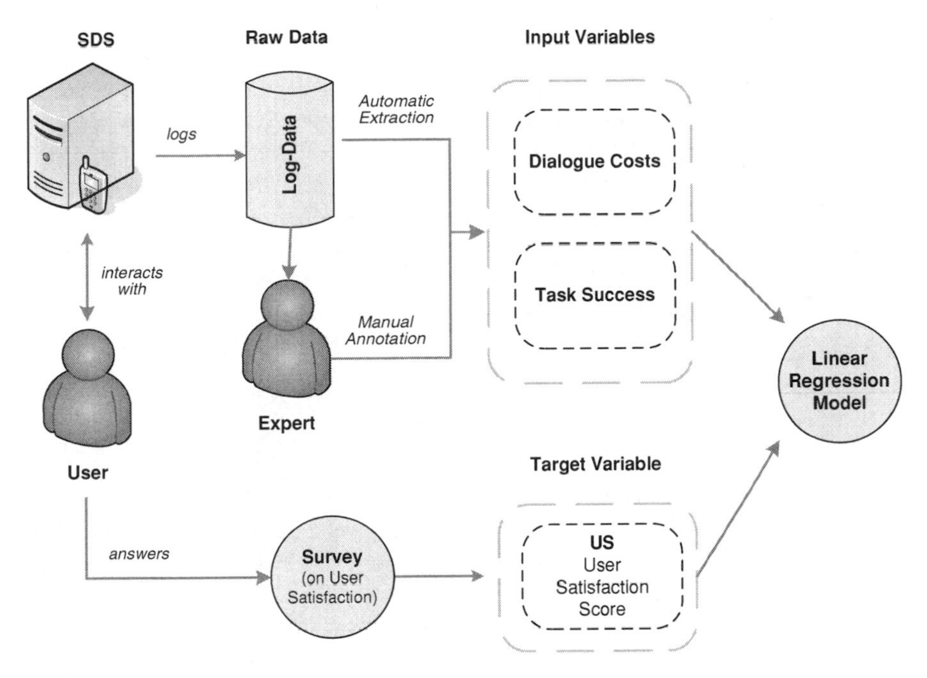

Fig. 2.11 The PARADISE framework for modelling the performance of an SDS (adapted from Walker et al. (2000))

It has been widely discussed in the recent years, see e.g. Möller (2005b), Hajdinjak and Mihelic (2006, 2007), Engelbrecht et al. (2008).

PARADISE targets three core goals:

- A comparison between multiple systems and multiple system versions from the same domains.
- The generation of prediction models for system usability based on system properties.
- The identification of properties having an impact on system usability.

The paradigm assumes that task success and dialog costs contribute to user satisfaction, which is the target variable in the model, see Fig. 2.11.

Hereby an automatic estimation of user satisfaction without the need to conduct extensive user tests should be enabled. The model is based on the assumption that user satisfaction is an indicator for the overall system performance Walker et al. (1998). It postulates that systems with high task success rates and low dialog costs yield high user satisfaction ratings implying a high system performance. PARADISE thus models user satisfaction as a multivariate linear regression problem.

Task Success

In PARADISE, task success is measured as a continuous variable with the κ coefficient (Walker et al. 1997). Average task success in the latter definition is measured on Attribute Value Matrix (AVM) instantiations from a set of dialogs for a specific scenario, e.g. "book a flight from Hamburg to Munich at 10am". The AVM instantiations contain an ordered set of pairs of information slots for the specific scenario and the values obtained from the user (e.g. departure_city='Berlin', arrival_city='Munich', dep_time='10am'). A confusion matrix is determined based on a set of different scenarios by comparing all user AVM instantiations of a scenario with the key AVM containing the intended correct values of this scenario. The kappa value is then obtained by calculating

$$\kappa_C = \frac{p_A - p_E}{1 - p_E}, \tag{2.43}$$

where p_A is the proportion of times the AVM instantiations for a set of dialogs agree with the key AVMs of the scenarios. p_E is the expected agreement by chance that the AVM instantiations map the key AVMs (Walker et al. 1998).

For larger systems with extensive dialog structures the implementation of task success as a κ coefficient appears complex and less suitable. Möller states that "the κ coefficient of task success (from PARADISE) does not adequately describe the quality of the task solution" (Möller 2005b). Other studies, e.g. Walker et al. (2000) use binary or teniary variables to measure task success (TS) (Danieli and Gerbino 1995) which is obtained from user surveys.

Dialog Costs

PARADISE uses a set of parameters to assess the costs of a dialog. Usually they are obtained by post processing the system logs or by manual annotation. Widely used variables to measure "dialog efficiency" in PARADISE are the number of turns or elapsed time until task completion. "Dialog quality" is determined as a mean recognition score of the ASR module, the number of help and cancel requests as well as the number of ASR rejections, time-out prompts and barge-ins (Walker et al. 1997, 2000).

User Satisfaction

The value of the target variable "user satisfaction" is obtained from user questionnaires querying central aspects that affect user satisfaction. Among others they comprise questions about perceived TTS and ASR performance, task ease, interaction pace, user expertise and future use (Walker et al. 2000). Answers are given on five-point scales that are summed to a final user satisfaction score ranging from 8 to 40.

A regressive user satisfaction model $\widehat{\mathrm{US}_w}$ can be determined with the input variables from task success, dialog efficiency and dialog costs:

$$\widehat{\mathrm{US}_w} = \alpha \cdot \mathscr{N}(\kappa) - \sum_{i=1}^{n} w_i \cdot \mathscr{N}(c_i), \tag{2.44}$$

where α is the weight for κ, the cost measures c_i are weighted by factors w_i. $\mathscr{N}$ is the Z-score normalization function (Cohen 1995a) that is employed to overcome differences between the varying scales of the cost values and κ.

Walker et al. applied PARADISE on data from two SDS at AT&T, the e-mail reader ELVIS and the train timetable system TOOT, and derived a joint model for system performance (Walker et al. 1998):

$$\widehat{\mathrm{US}_w} = 0.23 \cdot \mathrm{COMP} + 0.43 \cdot \mathrm{MRS} - 0.21 \cdot \mathrm{ET}, \tag{2.45}$$

where COMP is the subjective estimation of users on task success, which differs from κ and TS, MRS the mean recognition score and ET the elapsed time. All variables are Z-score normalized.

To a quite limited degree, a generalization of the model seems possible. Cross-domain and cross-system generalization at least between systems from the same laboratory has been explored in Walker et al. (2000) and Möller et al. (2008). The model performance drops significantly when unseen data from other systems is applied.

Möller et al. (2008) collected data from three interaction experiments by using two different systems. Test participants followed predefined scenarios and at the end they were asked about various aspects of the system, including the perceived overall quality. All experiments used a transcribing wizard instead of speech recognition to avoid an excessive impact of a speech recognition unit that was not yet optimized. The systems under analysis were the BoRIS restaurant information system and the INSPIRE smart-home systems. From the recorded dialogs, interaction parameters were extracted. Additionally to the multivariate linear regression as proposed within PARADISE, Möller et al. employed classification and regression trees (CART) and multilayer perceptrons (MLP) and examined the impact of different parameter sets. The parameters under analysis included subsets from ITU (2005) that lists parameters describing the interaction with spoken dialog systems, as well as parameters defined in PARADISE. They observed no significant differences in the prediction accuracy when applying linear and non-linear classifiers suggesting that it is irrelevant for the model performance whether a linear or non-linear relationship between input and target variable is assumed. The study further shows that prediction performances heavily decrease from seen to unseen data. For a detailed discussion on PARADISE refer to Möller (2005b).

2.3.1.2 Further Automated Evaluation Techniques

Other approaches exist that allow an estimation of system performance and quality that are shortly addressed in the following.

PARADISE-Related Data-Driven Studies

While the PARADISE model is trained with real user satisfaction scores from surveys, Evanini et al. pursue a recent approach that is entirely based on expert listeners (Evanini et al. 2008). Their main criticism about PARADISE is that real users may interpret the questions from the user survey differently while expert raters are expected to deliver a more stable rating when evaluating the performance of a dialog system. They further criticize the quality of the questions on the survey, which, in their opinion, lacks empirical evidence. A central difficulty in asking real users about their satisfaction in commercial SDS is the fact that surveys can only be placed at the end of successful dialogs. This would lead to a bias towards satisfied users. In their study, a set of 15 expert listeners rated subjectively the quality (on a scale from 1 to 5) of approximately 100 dialogs originating from a call routing system. 1,180 calls alongside with approximately three expert ratings per call have been chosen to train a decision tree. The caller experience ratings of the expert raters served as target variable of the model while the classification status of the call, i.e. how well the system determined the reason for the call, the task success as well as the number of ASR errors and operator requests served as input variables. A disjoint set of 202 test calls was used to evaluate the model. It could be shown that, despite its simplicity, the model revealed a similar performance as the human raters. The model is able to predict the ratings of the three expert listeners with 91.9 % accuracy provided that the automated rating differed one point at the maximum from the expert rating. In a similar spirit, Zweig et al. (2006) presented an automated call quality monitoring system that assigns quality scores to recorded calls based on speech recognition. However, the system is restricted to human-human conversation and the aim is to survey, whether operators behave courteously and appropriately in dealing with customers.

User Simulation

Eckert et al. (1998) proposed an evaluation technique that is based on an automated user. By that, a personal user bias and annotation errors are excluded. The user simulation is derived from an existing corpus with human-machine interactions. López-Cózar et al. (2002, 2003, 2006) proposed a "user simulator" that is derived from real interactions from users interacting with the SAPLEN system (López-Cózar et al. 2002). The simulator interacts with the dialog system and generates test data. A general advantage of user simulators over interactions captured from real users is that effects of changes from system components can be studied right away during the design phase. A deployment phase is thus not required to obtain test data for evaluation purposes. User simulators are listed for the sake of completeness in the field of statistical evaluation techniques, but they do not have a direct relevance for this work.

2.3.2 *Online Estimation of Task Success and User Satisfaction*

The increasing complexity of SDS in all fields raises the need for an instant evaluation of ongoing system-user interaction, which we will denote as "online monitoring" techniques. To date the field of evaluating spoken dialog interaction on the exchange level, i.e. at an arbitrary point during the interaction, has only been sparsely examined. While data-driven evaluation techniques such as PARADISE estimate system performance based on data from *completed* dialogs (offline), they are not suited to monitor an ongoing interaction at any given point in the dialog, e.g. to predict system performance and system-user satisfaction. The main difference here compared to the works dealing with offline evaluation is that it is not the system itself that is evaluated and which is to be improved, but a specific interaction between user and system. It should be noted that the system in this context is fixed and cannot be changed. Another difference lies in the evaluation level: online monitoring requires modeling of the interaction on a fine-grained exchange level of a dialog while PARADISE and its derivatives operate on the dialog level. The latter frequently represent a dialog with average performance values, such as the average ASR confidence or the number of speech recognition errors, leading to a manageable number of input variables. In contrast to this, online monitoring approaches have to deal with an increasing number of input variables that represent the previous interaction. As e.g. in Walker et al. (2002) , Kim (2007) and Levin and Pieraccini (2006) the parameters captured at the exchange level are then subject to a feature vector representing the overall interaction of the user and the system up to a certain point in the dialog. The consequent requirements towards the model are considerably larger.

It seems to be important to note that the term "online monitoring" does not only cover the prediction of user satisfaction (e.g. as $\widehat{US_w}$ in PARADISE). Online monitoring is the permanent observation of ongoing system-user interactions with the aim to predict a variety of target variables that can be helpful to render the interaction between the system and the user more robust, user-friendly and target-aimed. Such variables may range from task success over user satisfaction to poor speech recognition performance.

Task Success

Some of the first models that accomplish online monitoring in SDS were proposed by Langkilde et al. (1999)and Walker et al. (2002) . Both employ RIPPER (Cohen 1995b, 1996), a rule-learning algorithm, to implement a "Problematic Dialog Predictor" forecasting the outcome of calls in the HMIHY (How May I Help You) call routing system by AT&T (Gorin et al. 1996, 1997). The classifier aims to determine, whether a call belongs to the class "problematic" or "not problematic". The term "problematic" in this context refers to calls where the task is not completed, an agent took over the call or the user suddenly hung up without any obvious reason. "Non-problematic" calls end up with a successful completion of the call. The prediction of the classifier is used to escalate to a human operator. If a call is considered as non-problematic, the interaction is continued without intervention. Due to the nature of HMIHY, the dialogs are quite short with not more than five dialog turns. The authors built one classification model based on static feature vectors consisting of features extracted out of the first dialog exchange, and another model based on a static vector based on features from the first and the second exchange. The first model achieved an accuracy of 72.3 % (94.4 % recall, 79.1 % precision) and the second model of 87.0 % (49.5 % recall, 79.7 % precision) respectively while the baseline for guessing the majority class lies by 64.0 %.

The studies from Langkilde et al. (1999), Walker et al. (2002) inspired further studies on predicting problematic dialog situations: Bosch et al. (2001) reports about online detection of communication problems on the exchange level by using RIPPER as classifier and the word hypothesis graph plus the last six question types as training material. If communication problems, i.e. misrecognitions, are detected, the authors propose to switch to a more constrained dialog strategy. Since users tend to speak intentionally loud and slow when facing recognition errors—a situation in which a conventional speech recognizer has no training—these authors propose the use of two speech recognizers in parallel in order to detect hyper-articulated speech more robustly. It should be noted that the aim is not escalation but adaptation. Levin and Pieraccini (2006) combined a classifier with various business models to arrive at a decision to escalate a caller depending upon expected cost savings. The target application is that of a technical support automated agent. Again a RIPPER-like rule-learner has been used. Paek and Horvitz (2004) considered the influence of an agent queue model on the call outcome and included the availability of human operators in the decision process. A rather simple yet quite effective approach has been published by Kim (2007). This call outcome classifier uses sequences of user prompts and achieves an accuracy of 83 % after five turns. However, a large number of available information sources such as ASR and LU performance scores are not taken into account which implies that a higher accuracy on this data set would be achievable.

User Satisfaction

Studies described in the previous section consider dialogs as "problematic" where the task success is endangered. However, the prediction of task success does not necessarily reflect the actual satisfaction of the user. Other studies thus consider user satisfaction as target variable and by that, aim to recognize poor dialog quality.

Engelbrecht et al. (2009) modeled user satisfaction as a process evolving over time. It is the first approach that captures the user's quality perception during an interaction. In the experiment, subjects were asked to interact with a Wizard-of-Oz derived from the BoRIS (Möller 2005a) restaurant information system. Each participant followed five dialogs, which have previously been defined following six predefined scripts. This resulted in equally long dialog transcripts for each scenario, so e.g. dialog scenario 1 from participant 1 and dialog scenario 1 from all other participants show the same length. The users were constrained to rate their satisfaction on a five-point scale with "bad", "poor", "fair", "good" and "excellent" after each dialog step. The interaction was stopped during the user voting. For each scenario a separate model was trained with data from 25 participants and evaluated with 17 disjoint participants using HMMs. MSE and Mean absolute error (MAE_{max}) have been used for evaluation. For unseen data from two specific scenarios, the model reached an average MSE of 0.086 and 0.0374.

Higashinaka et al. (2010a) picked up the idea from Engelbrecht et al. (2009) and created a model to predict turn-wise ratings that was evaluated on human-machine and human-human dialogs. The data employed was not spoken dialog but text dialogs from a chat system and a transcribed conversation between humans. Although the authors speak of modeling user satisfaction, the ratings originate from expert listeners that were asked to "feel" like the users rather than employing real users. Two raters listened to the recorded interactions and provided turn-wise scores from 1 to 7 on smoothness ("Smoothness of the conversation"), closeness ("Closeness perceived by the user towards the system") and willingness ("willingness to continue the conversation"). Since expert raters provide the rating it is highly questionable whether the model indeed predicts user satisfaction until empirical evidence of the congruence of expert ratings and user ratings is provided. The issue of expert ratings versus user ratings is discussed in Sect. 5.1.

Hara et al. (2010) created n-gram models to predict user satisfaction based on 518 dialogs from real users interacting with a music retrieval system. The model is based on *overall* ratings from the users measuring their satisfaction on a five-point scale *after* the interaction. The authors calculated n-gram sequences of dialog acts and automatically labeled them with the final user rating. For each of the five possible satisfaction ratings plus one "task not completed" label an n-gram model was calculated resulting in six models. The 3-gram model achieved the best results with 34 % accuracy in distinguishing between the six classes at any point in the dialog. Under closer consideration of the confusion matrix (Table4 in Hara et al. (2010)) the prediction of turn-based user satisfaction scores given only one overall score seems hardly possible and is close to random: the prediction of the five user satisfaction classes reach an average f_1-score as low as 0.252, which is only 0.052 points above the random baseline of 0.2.

Not surprisingly there exists a large difference in model performance depending on the fact whether the model has been trained with dialog-level ratings or with turn-level ratings. A similar result as Hara et al. (2010) was obtained by Higashinaka et al. (2010b). They used HMMs trained on dialog-level ratings and compared them to the performance of HMMs trained on turn-level ratings. The performance of the dialog-level rating models was closer to random than to the values obtained when training the models with turn-level ratings.

Engelbrecht et al. (2009), Hara et al. (2010) and Higashinaka et al. (2010b) share the idea to model user satisfaction on the exchange-level with statistical approaches. The main differences concern the acquisition of the class label, i.e. whether real users or expert raters determined the "satisfaction"—and if the labels have been assigned on the exchange-level, or if they have been interpolated from dialog-level ratings.

The strength of the study from Engelbrecht et al. (2009) is that real users are employed. It is the first attempt to automatically predict user satisfaction ratings on the exchange-level. Although not addressed in the publication, the approach could be used to analyze and uncover interaction patterns that lead to positive or negative user votings. This would enable dialog system developers to uncover weaknesses in the dialog design. However, the study lacks a general, system-wide model that can be employed to achieve this. Further it is restricted to specific, predefined scenarios which are not cross-evaluated. In contrast, Higashinaka et al. (2010a) created a system-wide model that can be employed to predict user satisfaction at arbitrary points in the dialog. Crucial for a successful recognition of user satisfaction is the choice and appropriateness of the input variables. Higashinaka et al. (2010a, b) and Hara et al. (2010) employ a—mostly hand annotated—"dialog act" feature to predict the target variable user satisfaction. Engelbrecht et al. (2009) additionally employed contextual appropriateness, confirmation strategy and task success, most of which need hand annotation. Yet it is mandatory for an automatic prediction of user satisfaction to design and derive completely automatic features that do not require manual intervention.

Higashinaka et al. (2010a) use dialogs in written form that originate from text chat interfaces rather than spoken dialog interaction. It seems to be difficult to transfer these results on spoken dialogs.

2.4 Summary and Discussion

Summary

In Sect. 2.1 we have introduced supervised learning techniques that may be applied for a data-driven online monitoring of SDS. The presented techniques included KNN, Support Vector Machines, Linear Regression and HMMs. We have further seen how the performance of such classifiers is assessed with a large set of performance metrics.

We consider emotion recognition as one important aspect for online monitoring SDS. Hence, we have outlined principles and related work in speech-based emotion recognition in Sect. 2.2. Emotion theories, relevant paralinguistic and linguistic features, existing corpora and labeling techniques have been presented.

That followed, we have analyzed existing work related to data-driven evaluation and monitoring of SDS, see Sect. 2.3. These data-driven evaluation techniques, e.g. PARADISE, aim to statistically model user satisfaction or similar target variables, such as expert ratings, on completed dialogs. They operate on the dialog level, i.e. a set of input variables mirroring the interaction of system and user is mapped to a target variable representing the satisfaction of the user or the expert opinion with the dialog system. This is accomplished with linear regression models or decision trees and allows the estimation of user satisfaction with unseen data. The rising complexity of SDS and the request for more user-friendly, adaptive, naturally interacting systems augment the demand for porting such approaches to the exchange level. This allows an evaluation, i.e. monitoring, of the interaction at every single dialog step during the interaction. For the exchange level, some approaches exist that predict task success and estimate user satisfaction. Related work in that field has also been presented.

Discussion

The book at hand presents new statistical modeling approaches and thereby closes a number of gaps in the field of adaptivity and statistical modeling of poor ongoing human-machine interaction in SDS. To date, no research has been conducted that analyzes this problem in its entirety taking into account the user's emotional state and further interaction patterns that indicate task failure and low Interaction Quality at the same time.

Specifically the emotional state of the user has so far not been taken into account in this context, although emotions seem to have a major impact on task completion and could serve as indicator for user satisfaction. Particular issues with existing emotion recognition approaches are addressed in this work. A large proportion of studies dealing with emotion recognition rely on acted or laboratory speech, e.g. (Pittermann et al. 2009). Consequently, the speech samples stem from studio recordings or TV shows with high speech quality and noise-free audio data of substantial length. The results obtained can hardly be generalized on the recognition of real-life emotions. Unlike the conditions that are given when recognizing emotions in such high-quality monologues or human-human conversations, emotion recognition for SDS has to deal with a large number of speakers and with speech samples containing strong background noise, such as street traffic or subways. Furthermore, it has to deal with crosstalk and very short, command-based user utterances. Models created under laboratory conditions fail when applied on non-acted, non-laboratory speech. While some studies deal with real-life speech, further investigations are required to better model emotions in such settings. A major shortcoming hereby is the lack of sufficient real-life corpora that may be used for analysis. In the course of this book, a comprehensive data set of field interactions is annotated resulting in a large emotion corpus. This corpus allows an analysis of the dependency between system design, system events and interaction patterns. A dependency of these variables with the emergence of emotions can be assumed. More recent work that has dealt with real-life speech neglects these knowledge sources that can be derived from the overall interaction and employs merely acoustic/prosodic and linguistic information. We thus enhance acoustic and linguistic modeling with Interaction Quality-related modeling and analyze the impact of these interaction-related features to emotion recognition. Until today it has further not been analyzed how specific emotions evolve during an interaction and which factors encourage their occurrence. Related studies consider solely the recognition of isolated utterances rather than taking into account the context of an utterance when classifying. Beyond, the performance metrics of related emotion studies are frequently insufficient and allow us to only read little into the performance of emotion recognition in field conditions. Statistical classifiers and models that deliver a high performance on a well-defined training set may completely fail when deployed in field scenarios where the distribution of emotional speech is a completely different one.

In the second instance, this book addresses further issues related to critical dialog situations and targets a prediction of task success and poor Interaction Quality based on interaction patterns. Related work in this field mainly focused on estimating system quality on the dialog level (Walker et al. 1997, 1998, 2000; Hajdinjak and Mihelic 2006; Möller et al. 2008) using data-driven approaches. These allow simply a system-wide evaluation rather than an evaluation of a specific dialog scenario.

For the dialog level, a comprehensive set of domain-independent interaction parameters exists that models entire dialogs. This set is described in the ITU-T Recommendation 24 to P-Series (ITU 2005). For the novel approaches presented in this work, parameters are required that describe the interaction in spoken dialogs on the exchange level. A complementary set of interaction parameters in this style is herefore introduced. It can be entirely derived from system logs. The central strength of this approach is the absence of manual annotation, which eventually allows a deployment of the related models in real-life applications.

The evaluation of specific dialog interactions on the exchange level, i.e. during an interaction, has played a very subordinate role so far. While existing related work operating on this exchange level aims to predict, whether a task between the system and the user is likely to be completed, they do not exploit all available information that may automatically be derived from the interaction. Instead, they restrict the input variables to a very limited set. Secondly, the existing approaches are predominantly not suited for long-lasting spoken dialog interactions, as e.g. Walker et al. (2002) who aim to predict the task success only in the first two exchanges. Newer generations of SDS allow dialogs of substantial length and the present techniques have not been evaluated for such scenarios.

Further work related in that field addresses the prediction of user satisfaction on the exchange level. A number of issues are solved in the book at hand. Similarly as in related work on task success, studies in predicting user satisfaction limit the estimation of user satisfaction on a very restricted set of input variables. In most work, only a hand-annotated dialog act feature is employed that apparently does not model the problem reliably. This book explores the performance contribution of the new input variables for the task of dialog quality and user satisfaction estimation. Related work does furthermore neglect the emotional state as input feature, while at the same time, this could give information on the satisfaction of the user. The contribution of emotion recognition for that task is examined. Exchange-level labels of quality are required for predicting the quality in SDS on exchange-level. However, capturing satisfaction scores from users in field-scenarios during an interaction is virtually not reasonably viable. On top of this, user studies are time-consuming and costly. Capturing user satisfaction in laboratory conditions experiences similar difficulties and poses likewise the question, how the actual satisfaction of a user can be reliably tracked during the interaction. As we have seen in Sect. 2.3.2, related studies thus either capture user satisfaction in laboratory settings with a permanent interruption of the task or they completely relinquish the idea of tracking real user satisfaction and use expert annotations instead. In this context it is questionable whether expert ratings indeed mirror the actual satisfaction of the user. This book follows a pragmatic approach to solve this issue. We propose the term "Interaction Quality" to assess the quality of the interaction from an expert's point of view. In the course of this book, a number of expert annotators annotated real-life dialogs and by that, establish an objective quality model for evaluating SDS interaction on the exchange level. The correlation of user satisfaction with this IQ metric is hereupon evaluated based on a study with real user satisfaction scores obtained from a lab study with users interacting with the same system. The central advantage of the IQ metric is that it can be ported to dialog systems of arbitrary domains and is easily reproducible without conducting user studies.

Chapter 3
Interaction Modeling and Platform Development

In the following we describe data sets, parameters, and techniques that are pivotal to all the following chapters serving as basis for the proposed machine learning approaches. The majority of novel techniques presented in this book are developed and evaluated on data sets from applications containing interactions from real users performing real tasks. They originate from commercial and non-commercial systems serving a large number of users throughout the United States and are presented in Sect. 3.1. Raw data sets respectively comprising several hundreds to several tens of thousands of dialogs has been transformed, enriched with features, transcribed and annotated by means of human annotators and by that prepared for machine learning tasks. In Sect. 3.2 we address the parameterization and annotation of the presented raw data. For this purpose a large set of interaction parameters is introduced. It serves as input variables for the majority of the presented statistical models. Ensuing, the labeling process including annotation schemes for negative emotions is presented and discussed. The distribution of negative emotions in real-life interactions is comprehensively analyzed.

The data-driven, statistical models intended to predict critical dialog situations and user characteristics require a platform where logged dialogs can be managed, analyzed, and annotated. Such a "workbench" is presented in Sect. 3.3. The platform supports the handling of large dialog corpora as well as the development and evaluation of data-driven models.

3.1 Raw Data

The main data sets that have been predominantly employed in this book originate from three different domains:

- LevelOne Broadband Agent (SpeechCycle, Inc.): a commercial, automated technical support agent helping callers to recover lost passwords and fix lost or slow

A. Schmitt and W. Minker, *Towards Adaptive Spoken Dialog Systems*,
DOI: 10.1007/978-1-4614-4593-7_3,

Table 3.1 Characteristics of the real-life corpora employed in the book: SpeechCycle LevelOne Broadband Agent, SpeechCycle LevelOne Video Agent, and the CMU Let's Go Bus Information System (Field + Lab)

	SpeechCycle BroadBand Agent	SpeechCycle VideoAgent	CMU Let's Go Field	CMU Let's Go Lab
	SC-BAC	SC-VAC	LG-FIELD	LG-LAB
# Dialogs	104,800	29,269	328	140
Incl. user audio	1,911	328	200	140
Incl. full audio	–	328	200	–
Type	problem solving	problem solving	information retrieval	information retrieval
Domain	Internet	TV/VOD	bus schedules	bus schedules
Language	US-American	US-American	US-American	non-native English
Avg. # exchanges	26.9(±20.8)	18(±10.63)	26(±21.5)	22.1(±10.3)
Avg. duration/s	297.2(±306.1)	198.3(±219.65)	116.8(±114.1)	135.4(±70.5)
Human assistance	live agent	live agent	–	–

The LevelOne systems allow the users to opt for a human operator. Moreover, they escalate users to operators in case the problem to be solved is out of scope. In the CMU Let's Go System such an escalation is not provided

Internet connections. It is deployed in different variations for large Internet providers throughout the United States.

- LevelOne Video Agent (SpeechCycle, Inc.): a commercial, automated technical support agent helping callers to solve problems related to video on demand (VOD) and television services. The VideoAgent is running in different versions for large cable TV providers and serves customers across the United States.
- Let's Go Bus information system Carnegie Mellon University (CMU): Let's Go delivers bus schedule information to citizens of the city of Pittsburgh. It was created at CMU and answers 40–60 calls a day (in 2006).

For details of the systems refer to Acomb et al. (2007) for the Broadband and Video Agent and Raux et al. (2006) for the Let's Go system. Three corpora have been derived from raw data provided by SpeechCycle and the CMU. The fourth corpus, the Let's Go Lab corpus, has been created based on collected raw data from user experiments under laboratory conditions. Details of this process are described in Sect. 5.3. The characteristics of the data are depicted in Table 3.1.

Figure 3.1 shows the distribution of the call duration of the single corpora. Many calls from all three systems last less than 100 s. While for Let's Go this lies in the nature of a bus schedule information system the Video- and Broadband Agents determine the user's reason of the call in one of the first dialog steps. For categories where no automated troubleshooting is implemented, the customers are directly routed to a human expert operator. This explains the large number of short dialogs. The remaining calls are subject to automated troubleshooting by the dialog system and exhibit frequently durations of 10 min and more.

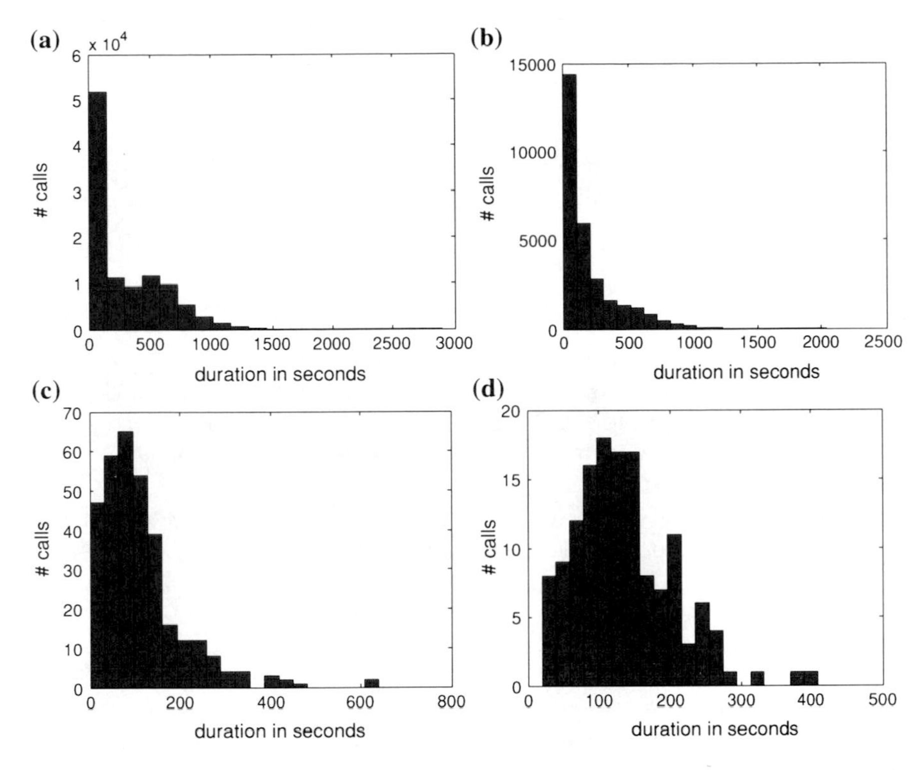

Fig. 3.1 Histograms of dialog durations from the Broadband Agent, Video Agent, and Let's Go systems. **a** Broadband Agent—SC-BAC. **b** Video Agent—SC-VAC. **c** Let's Go Field—LG-FIELD. **d** Let's Go Lab—LG-LAB

Users of the SpeechCycle LevelOne systems are customers from companies deploying the Broadband- and VOD service for technical support. The users call from home over land line or mobile connections and by that from a comparatively quiet environment without constant background noise. However, cross-talk, barking dogs, arguing children, bubble, etc. do occur. Further peculiarities of the respective applications exist. Users of the Broadband Agent are sometimes asked by the system to check cables or search for the modem. As a result, many customers switch to hands-free mode and communicate with the system remotely. Users from the Let's Go system call mostly in the evening, at night or in the early morning, when the phones of the bus company are not manned by human operators. Frequently the callers sound weary or drunk. Strong background noise from pubs, shopping mall, or street noise accompanies the speech, since a majority of the users connect to the system while on the move. The Let's Go system exhibits an obviously higher amount of non-native speakers compared to the SpeechCycle systems. The Speech-Cycle LevelOne systems handle several hundreds of thousand calls per year, whereas the Let's Go system answers several ten thousand per year (Raux et al. 2005).

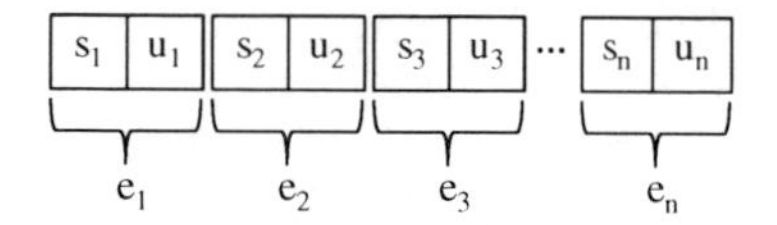

Fig. 3.2 Structure of a system-initiative, directed spoken dialog, as in SC-BAC, SC-VAC, LG-FIELD, and LG-LAB. The interaction is modeled on the exchange level *e* comprising data from system turn *s* and user turn *u*

3.2 Parameterization and Annotation

The data presented in this work have been used to implement statistical models that enable a prediction of various target variables online, i.e., while the dialog is going on, such as the emotional state of the user. For a statistical modeling, input variables and target variables have to be defined.

The raw data requires parameterization, i.e., a definition of the input variables, and annotation, i.e., the definition of the target variables.

3.2.1 Interaction Parameters

An interaction between the system and the user can be quantified by means of interaction parameters, which are derived from raw system logs. For example, such parameters may capture the number of misrecognitions in a dialog, the number of times the user did not respond in time, the duration or average number of words per user response. On the dialog level, a comprehensive set of such parameters has been defined in the ITU Supplement 24 to P Series "Parameters describing the interaction with spoken dialog systems" (ITU 2005). The set defines parameters that quantifiy an interaction over *completed* dialogs. To date, no strictly defined, comprehensive set of interaction parameters on the exchange level exists. Such a set will be required in the following for accomplishing online monitoring. For a better understanding of the modeling, the structure of a system-initiative, directed spoken dialog is depicted in Fig. 3.2.

Virtually all dialogs from systems in the field including the corpora described above follow this structure. The smallest unit in a dialog is the *turn*, which can either be a system turn *s* or a user turn *t*. Both together form an exchange *e* which spans our modeling unit. It should be noted that the term "exchange" is considered a technical term here and thus might not be in line with a linguistic definition. Due to technical reasons and the dialog design of a specific SDS it may also be possible that an exchange is limited to a single system turn, such as "One moment please, let me check the bus schedule for you...". Such a system turn does not expect a user response and is nevertheless modeled as "exchange". In the following, we define the parameters that can be derived from system logs originating from the system modules ASR, LU, and DM. They describe the interaction within an

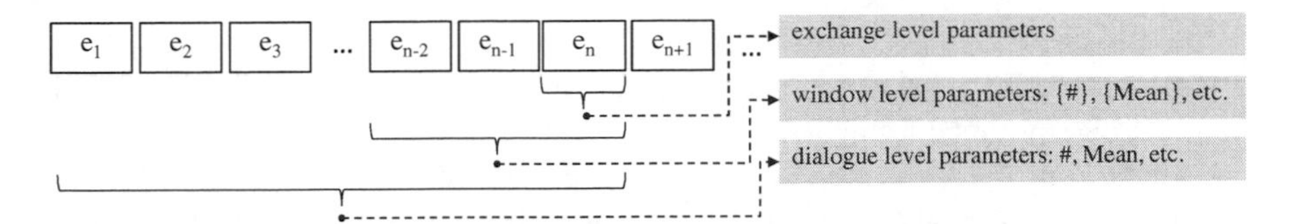

Fig. 3.3 The three different modeling levels representing the interaction at exchange e_n: the most detailed exchange level, comprising parameters of the current exchange; the window level, capturing important parameters from the previous n dialog steps (here $n = 3$); the dialog level, measuring overall performance values from the entire previous interaction

exchange plus the parameters that summarize important system variables beyond, i.e., prior, to the current exchange. To account for the overall history of important system events such as the ASR confidence, the answer behavior of the user, etc. we added running tallies counting important events plus percentages and mean values for certain features symbolized with the suffixes '#' (for running tallies), '%' and 'MEAN'. Parameters pertaining to the immediate context, i.e., a window of the current exchange respectively contain the suffixes '{MEAN}' such as the average confidence score of the ASR in the immediate context, and '{#}', e.g., for the number of times the user recently interrupted the system. The architecture is depicted in Fig. 3.3.

Some of the interaction parameters described in the following can be found informal and in excerpts in related works which are respectively cited. Most related studies employed the respective parameters for dialog level modeling (i.e., PARADISE and derivatives), so they are originally not intended for an interaction modeling on the exchange level and require a re-definition.

Automatic Speech Recognition Parameters

The Automatic Speech Recognition (ASR) parameters describe performance values and variables from the automatic speech recognition module (cf. pipeline architecture in Fig. 1.2). They summarize how reliable the ASR recognizes the spoken user utterance. Further, they track if the user interrupted or did not respond to a system question. Finally, they track the modality the user chose to interact with the system. The related parameters are summarized in Table A.1 and described in the following.

The names of all activated grammars represented as a string are captured in the parameter GRAMMAR. The activated grammars indicate the possible and allowed user input, e.g., "DTMFYesNo.grxml; DTMFoperator.grxml; ynOnly.xml; operator.xml;" (Langkilde et al. 1999; Walker et al. 2002; Litman et al. 2000). The name of the grammar that returned the speech recognition parse is captured in TRIGGERED-GRAMMAR. The name of the triggered grammar might be an indicator for the interaction the user chose (Langkilde et al. 1999; Walker et al. 2002; Riccardi and Gorin 2000).

The UTTERANCE parameter represents the ASR parse from the user utterance, i.e., the automatically transcribed text string from the speech recognizer. It should be noted that throughout this work, the UTTERANCE parameter is considered as an entire text string. Alternatively, a bag-of-words realization (Albalate and Minker 2011) could be taken into account. The bag-of-words approach expands the feature space to n dimensions, where n is the number of all valid words defined in the ASR grammars. Instead of considering the entire utterance, the number of occurrences of a word is then marked in the bag-of-words vector. Given sufficient amounts of data, this approach might be more accurate compared to a use of the entire text string, but requires a substantial amount of data to work reliably. Vector representations of text strings are also discussed in Sect. 2.2.4. The utterance is likewise employed in its entirety as input feature in Langkilde et al. (1999), Walker et al. (2002) and Levin and Pieraccini (2006). It should be noted that utterances captured in IVR systems are of very short nature and mostly do not exceed a WPUT of 2–3 words. Vector model representations of the utterance—in contrast to taking the entire utterance as is—will thus most likely not improve the statistical prediction model.

The status from the speech recognition system is captured in the parameter ASRRECOGNITIONSTATUS. It represents the outcome of the ASR when trying to parse the user input. It can be one of {success, reject, timeout}. "success" refers to a—from the ASR's point-of-view—successful recognition of the utterance, i.e., the decoded word string matches an active grammar, "reject" means the ASR could not recognize the utterance and did not find a corresponding word sequence according to the active grammars; "timeout" indicates that the user did not respond within a given time slot (Langkilde et al. 1999; Walker et al. 2002; Levin and Pieraccini 2006).

Furthermore, the parameter #ASRSUCCESS counts the numbers of successfully parsed turns up to exchange e_n and can be defined as

$$\#\text{ASRSUCCESS} = \sum_{i=1}^{n} x \begin{cases} 1 & \text{ASRRECOGNITIONSTATUS} = \text{success}, \\ 0 & \text{otherwise}, \end{cases} \tag{3.1}$$

(Paek and Horvitz 2004; ITU 2005).

To measure the successful recognition attempts of the SDS in the immediate context of the current exchange e_n, the {#}ASRSUCCESS parameter is introduced. It counts the numbers of successfully parsed turns within the previous w turns prior to e_n, where w is the size of the window and can be calculated as follows:

$$\{\#\}\text{ASRSUCCESS} = \sum_{i=n-w}^{n} x \begin{cases} 1 & \text{ASRRECOGNITIONSTATUS} = \text{success} \\ 0 & \text{otherwise.} \end{cases} \tag{3.2}$$

The overall success rate of the ASR module is captured as % ASRSUCCESS, which tracks the percentage of "success" turns in all previous exchanges:

$$\{\%\}\text{ASRSuccess} = \frac{1}{n}\sum_{i=1}^{n} x \begin{cases} 1 & \text{ASRRecognitionStatus} = \text{success} \vee \\ & \text{ActivityType} \notin \{\text{question, confirmation}\} \\ 0 & \text{otherwise,} \end{cases} \tag{3.3}$$

(Paek and Horvitz 2004; Litman et al. 1999; Litman and Pan 2002; ITU 2005).

A time-out event is observed when the user did not deliver a response within a certain time slot after a system prompt. The duration of the time slot is system-dependent. In case the user does not respond in time, the ASR delivers a time-out, also called "no-input" event. It can be assumed that a high number of time-out events indicate a user that is not familiar with dialog systems or is distracted by his environment. Time-out-related parameters can thus be considered as very relevant for tracking critical dialog situations. The number of time-outs in the entire dialog is tracked with the parameter #TimeoutPrompts, which is the number of time-out events up to the current exchange e_n. It can be calculated as follows:

$$\#\text{TimeoutPrompts} = \sum_{i=1}^{n} x \begin{cases} 1 & \text{ASRRecognitionStatus} = \text{time-out,} \\ 0 & \text{otherwise,} \end{cases} \tag{3.4}$$

(Paek and Horvitz 2004; Kim 2007; Kamm et al. 1998; Litman and Pan 1999; ITU 2005).

Time-outs in the immediate context of the current exchange e_n are tracked with {#}TimeoutPrompts, which are the number of time-out turns within the previous w turns prior to e_n, where w is the size of the window.

$$\{\#\}\text{TimeoutPrompts} = \sum_{i=n-w}^{n} x \begin{cases} 1 & \text{ASRRecognitionStatus} = \text{time-out,} \\ 0 & \text{otherwise.} \end{cases} \tag{3.5}$$

The rate that represents the response behavior of the user is captured with % TimeoutPrompts, which is the percentage of "time-out" turns in all previous exchanges:

$$\%\text{TimeoutPrompts} = \frac{1}{n}\sum_{i=1}^{n} x \begin{cases} 1 & \text{ASRRecognitionStatus} = \text{time-out} \\ 0 & \text{otherwise,} \end{cases} \tag{3.6}$$

(Paek and Horvitz 2004; Litman et al. 1999; Walker et al. 2000).

To keep track of the recognition performance of the ASR, we track the number of ASR rejections up to the current exchange e_n as #ASRRejections, which can be determined as follows:

$$\#\text{ASRRejections} = \sum_{i=1}^{n} x \begin{cases} 1 & \text{ASRRecognitionStatus} = \text{reject,} \\ 0 & \text{otherwise,} \end{cases} \tag{3.7}$$

(Paek and Horvitz 2004; Levin and Pieraccini 2006; Kamm et al. 1998; Litman et al. 1999; Litman and Pan 1999; Walker et al. 2000; ITU 2005).

The misrecognitions in the immediate context of the current turn are captured with the parameter {#}ASRREJECTIONS, which is the number of ASR rejections within the previous w turns prior to e_n, where w is the size of the window.

$$\{\#\}\text{ASRREJECTIONS} = \sum_{i=n-w}^{n} x \begin{cases} 1 & \text{ASRRECOGNITIONSTATUS} = \text{reject}, \\ 0 & \text{otherwise.} \end{cases} \tag{3.8}$$

The interaction pace is heavily influenced by the functioning of the ASR module. ASR rejections and ASR time-out events are thus summarized in joint parameters to track malfunctioning ASR. The #TIMEOUTS_ASRREJ parameter counts the number of time-out and ASR rejection events up to the current exchange e_n. It can be defined as follows:

$$\#\text{TIMEOUTS\_ASRREJ} = \sum_{i=1}^{n} x \begin{cases} 1 & \text{ASRRECOGNITIONSTATUS} \in \{\text{time-out, reject}\}, \\ 0 & \text{otherwise.} \end{cases} \tag{3.9}$$

The parameter {#}TIMEOUTS_ASRREJ

$$\{\#\}\text{TIMEOUTS\_ASRREJ} = \sum_{i=n-w}^{n} x \begin{cases} 1 & \text{ASRRECOGNITIONSTATUS} \in \{\text{time-out, reject}\}, \\ 0 & \text{otherwise} \end{cases} \tag{3.10}$$

tracks the number of time-out and ASR rejection events within the previous w turns prior to e_n, where w is the size of the window.

The average rate of non-successful recognition attempts is summarized with the parameter %TIMEOUTS_ASRREJ, which is the percentage of time-out and ASR rejection events in all previous exchanges:

$$\%\text{TIMEOUTS\_ASRREJ} = \frac{1}{n} \sum_{i=1}^{n} x \begin{cases} 1 & \text{ASRRECOGNITIONSTATUS} \in \{\text{timeout,reject}\} \\ 0 & \text{otherwise.} \end{cases} \tag{3.11}$$

Exchanges in the beginning of the dialog where no ASR is active have to be replenished with "0".

Speech-events from the user that interrupt the system-prompt are marked with BARGE- IN?, which is "true" if the user interrupted the system prompt, and "false" otherwise ($\in$ {true,false}). "Barge-in" is originally intended for expert users as short-cut for long system prompts (McTear 2004). By using barge-in, the user is able to speed up the interaction with the system. While the presence of barge-in can be an indicator for expert users, it can also indicate impatient users or users that do not operate in quiet environments and who cause unintentional interruptions.

We track the number of barge-ins up to the current exchange e_n as variable #BARGE- INS:

$$\#\ \text{BARGE- INS} = \sum_{i=1}^{n} x \begin{cases} 1 & \text{BARGEIN?} = \text{true}, \\ 0 & \text{BARGEIN?} = \text{false}, \end{cases} \tag{3.12}$$

(Kamm et al. 1998; Litman et al. 1999; Litman and Pan 1999; Walker et al. 2000; ITU 2005).

We further track the barge-ins within the immediate context within the previous w turns prior to e_n, where w is the size of the window:

$$\{\#\}\ \text{BARGE- INS} = \sum_{i=n-w}^{n} x \begin{cases} 1 & \text{BARGEIN?} = \text{true}, \\ 0 & \text{BARGEIN?} = \text{false}, \end{cases}$$

as well as the barge-in rate of the entire interaction as %BARGE- INS:

$$\%\ \text{BARGE- INS} = \frac{1}{n} \sum_{i=1}^{n} x \begin{cases} 1 & \text{BARGEIN?} = \text{true}, \\ 0 & \text{BARGEIN?} = \text{false}, \end{cases} \tag{3.13}$$

which is the percentage of barge-ins in all previous exchanges (Litman et al. 1999; Walker et al. 2000).

We further parameterize the confidence of the ASR module representing the certainty of returning the correct ASR parse as ASRCONFIDENCE parameter ($\in \mathbb{R}\{0..1\}$) (Langkilde et al. 1999; Paek and Horvitz 2004; Levin and Pieraccini 2006; Kamm et al. 1998; Walker et al. 2000; Litman and Pan 2002).

As overall understandability metric we apply the MEANASRCONFIDENCE, which is the average ASR confidence up to this exchange in the entire interaction (Litman et al. 1999; Litman and Pan 1999):

$$\begin{aligned} &\text{MEANASRCONFIDENCE} \\ &= \frac{1}{n} \sum_{i=1}^{n} x \begin{cases} \text{ASRConfidence}_i & \text{ASRRECOGNITIONSTATUS} \\ & \in \{\text{success, reject}\} \\ \text{ASRConfidence}_{\text{corpus\_mean}} & \text{otherwise.} \end{cases} \end{aligned} \tag{3.14}$$

A low MEANASRCONFIDENCE value indicates that the system has difficulties in understanding the user. It is easy to comprehend that this value is related to the number of misrecognitions #ASRREJECTIONS. Missing values are replenished with the average corpus-wide confidence $\text{ASRConfidence}_{\text{corpus\_mean}}$ based on the average confidence of all "success" and "reject" turns.

The understandability of the user in the immediate context is tracked as {MEAN} ASRCONFIDENCE parameter:

$$\{\textsc{Mean}\}\textsc{ASRConfidence} = \frac{1}{n}\sum_{i=n-w}^{n} x \begin{cases} \text{ASRConfidence}_i & \textsc{ASRRec.Status} \\ & \in \{\text{success, reject}\} \\ \text{ASRConfidence}_{\text{corpus\_mean}} & \text{otherwise,} \end{cases} \quad (3.15)$$

which is the average ASR confidence within previous w turns prior to e_n, where w is the size of the window. Missing values are replenished with the average corpus-wide confidence $\text{ASRConfidence}_{\text{corpus\_mean}}$ based on the average confidence of all "success" and "reject" turns.

We further time the duration of the user utterance in seconds, user turn duration (UTD) (Langkilde et al. 1999; Walker et al. 2002; Levin and Pieraccini 2006; ITU 2005). For system prompts without user answer this value is set to zero.

Some dialog systems, such as IVR systems, allow the user to enter information using different modalities, such as voice, touchtone (i.e., DTMF), or both. The available choice of the input modality may differ from turn to turn. A common mixed-modality system prompt might be "For customer service press '1' or say 'Customer Service', ...". The expected modality that is explicitly offered to the user is tracked with the parameter ExMo. It is the expected input modality by the system at the current exchange ∈ {speech, dtmf, both, none}, see also Langkilde et al. (1999) and Walker et al. (2002).

The modality the user eventually employs for responding to the prompt is tracked with the parameter Modality, ∈ {speech, dtmf} (Langkilde et al. 1999; Walker et al. 2002; Levin and Pieraccini 2006). Signs for unexpected user behavior, which eventually might lead to critical dialog situations, can be the presence of unexpected modality usage. A user pressing DTMF keys although touchtone usage has not been offered in the current system prompt might give evidence that the user requires help or wants to bypass the system. The UnExMo? parameter tracks this behavior:

$$\textsc{UnExMo?} = \begin{cases} \text{true} & \textsc{Modality} \notin \textsc{ExMo} \\ \text{false} & \text{otherwise.} \end{cases} \quad (3.16)$$

The number of unexpected modality usages is tracked with #UnExMo, which is the number of unexpected modality usages up to this exchange e_n:

$$\#\textsc{UnExMo?} = \sum_{i=1}^{n} x \begin{cases} 1 & \textsc{Modality} \notin \textsc{ExMo} \\ 0 & \text{otherwise.} \end{cases} \quad (3.17)$$

We further define the parameter {#}UnExMo as the number of unexpected modality usages within previous w turns prior to e_n, where w is the size of the window:

$$\{\#\}\textsc{UnExMo?} = \sum_{i=n-w}^{n} x \begin{cases} 1 & \textsc{Modality} \notin \textsc{ExMo} \\ 0 & \text{otherwise,} \end{cases} \quad (3.18)$$

and finally the parameter %UNEXMO as percentage of unexpected modality usage up to this exchange:

$$\%\text{UNEXMO?} = \frac{1}{n}\sum_{i=1}^{n} x \begin{cases} 1 & \text{MODALITY} \notin \text{EXMO} \\ 0 & \text{otherwise.} \end{cases} \tag{3.19}$$

As the last variable that can be derived from the ASR module we introduce WPUT, the "words per user turn", which is the number of words returned in the ASR parse (Langkilde et al. 1999; Walker et al. 2002; Levin and Pieraccini 2006; ITU 2005).

(Spoken) Language Understanding Parameters

The second set of features is derived from the (spoken) language understanding (SLU) module (Fig. 1.2) that extracts the semantic meaning from the automatically transcribed user utterance. Special attention is drawn to two important events that can be derived from the semantic parse: the user's request for human assistance (operator request) and the request for more information (help request), which is either a demand for a more detailed explanation of the current system step or an expression of the user's helplessness during the interaction process. It should be noted that the operator request feature can only be collected if the system includes the possibility of human assistance. In systems where obviously no human operators stand by, such as in the Let's Go system, a request for operators are unlikely to be observed and can thus be considered as statistically irrelevant. The presented parameters are listed in Table A.2 in the appendix and are described in detail in the following.

The SEMANTICPARSE contains the dense semantic meaning of the caller utterance as returned by the activated grammar, cf. Litman et al. (1999). For example the semantic meaning of an utterance "I want to speak to a representative" could be parsed as "operator" with an appropriate grammar. Based on the semantic meaning, further information, e.g., about the user's need for help can be derived. With the parameter HELPREQUEST? we track whether the user explicitly asked for more detailed information to the current system turn, i.e., whether the current turn is a (from the system recognized) help request ($\in$ {true, false}).

With the parameter #HELPREQUESTS we count, how frequently the user has already asked for help during the interaction up to the current exchange e_n:

$$\#\text{HELPREQUESTS} = \sum_{i=1}^{n} x \begin{cases} 1 & \text{HELPREQUESTS?} = \text{true}, \\ 0 & \text{otherwise,} \end{cases} \tag{3.20}$$

(Paek and Horvitz 2004; Kamm et al. 1998; Litman and Pan 1999; ITU 2005; Hajdinjak and Mihelic 2006).

The number of help requests within the previous w turns prior to the current exchange e_n, where w is the size of the window, is tracked with the parameter {#}HELPREQUESTS:

$$\{\#\}\text{HELPREQUESTS} = \sum_{i=n-w}^{n} x \begin{cases} 1 & \text{HELPREQUESTS?} = \text{true}, \\ 0 & \text{otherwise}. \end{cases} \tag{3.21}$$

Finally, %HELPREQUESTS captures the percentage of help requests in all previous exchanges:

$$\%\text{HELPREQUESTS} = \frac{1}{n} \sum_{i=1}^{n} x \begin{cases} 1 & \text{HELPREQUESTS?} = \text{true} \\ 0 & \text{otherwise}, \end{cases} \tag{3.22}$$

(Paek and Horvitz 2004; Litman et al. 1999; Hajdinjak and Mihelic 2006).

The request for a live agent or human assistance can give some indication that the user wants to opt out or bypass the system. It is further an indicator for users that feel unable to cope with an SDS. Whether the currently observed exchange is a request for such a live agent is tracked with the variable OPERATORREQUEST? ($\in$ {true, false}). The information is retrieved from the SEMANTICPARSE and requires that the system correctly parses the user input, cf. Paek and Horvitz (2004). Furthermore, we count the number of operator requests as #OPERATORREQUESTS that occur up to the current exchange e_n:

$$\#\text{OPERATORREQUESTS} = \sum_{i=1}^{n} x \begin{cases} 1 & \text{OPERATORREQUEST?} = \text{true}, \\ 0 & \text{otherwise}, \end{cases} \tag{3.23}$$

(Paek and Horvitz 2004), as well as the number of operator requests {#}OPERATOR REQUESTS within the previous w turns prior to e_n, where w is the size of the window:

$$\{\#\}\text{OPERATORREQUESTS} = \sum_{i=n-w}^{n} x \begin{cases} 1 & \text{OPERATORREQUEST?} = \text{true}, \\ 0 & \text{otherwise}. \end{cases} \tag{3.24}$$

The average request for an operator in the ongoing dialog is determined with %OPERATORREQUESTS, which is the percentage of operator requests in all previous exchanges:

$$\%\text{OPERATORREQUESTS} = \frac{1}{n} \sum_{i=1}^{n} x \begin{cases} 1 & \text{OPERATORREQUEST?} = \text{true} \\ 0 & \text{otherwise}, \end{cases} \tag{3.25}$$

(Paek and Horvitz 2004)

Dialog Manager Parameters

As central unit in a dialog system the dialog manager (DM) (Fig. 1.2) keeps track of the interaction and provides information such as the current dialog step the system is in, the time elapsed since the beginning of the dialog, the number of tries to fill the semantic slot etc. The DM-related parameters are defined in the following and are comprehensively listed in Table A.3 in the appendix.

First, we track durational features and information that is related to the dialog progress. We capture the parameter #EXCHANGES by counting the number of system-user exchanges that have occurred up to the current exchange and differentiate furthermore between #SYSTEMTURNS and #USERTURNS. It should be noted that the number of exchanges #EXCHANGES may differ from the number of system or user exchanges since some activities may consist of a single user or system turn (Langkilde et al. 1999; Walker et al. 2002; Levin and Pieraccini 2006; Litman and Pan 1999; ITU 2005; Möller et al. 2008).

Furthermore, we keep track of the number of system questions #SYSTEMQUESTIONS that have been posed up to this exchange (ITU 2005) as well as the number of system questions in the current context window {#}SYSTEMQUESTIONS. Finally, the dialog duration DD in seconds up to the present exchange is captured (Langkilde et al. 1999; Walker et al. 2002, 2000; ITU 2005).

The ACTIVITY parameter keeps track of the activity that was performed by the system consisting of an identifier for the question or statement. Activities of different dialog systems are defined according to the flowchart design, e.g., a bus information service may have activities like "query.arrival_place", "query.travel_time", "confirm_okay" etc. The names are determined by the system designer and are system-dependent, see also Langkilde et al. (1999) and Walker et al. (2002). In a way they can be compared to domain-dependent, system dialog acts (Stolcke et al. 2000). The variable ACTIVITYTRIGRAM comprises a sequence of the current activity plus the two previous activities. This parameter models the history of activities and mirrors the recent dialog flow. We have chosen a size of "3" for the n-gram since it constitutes a tradeoff between spanning a reasonable history of the dialog while keeping the necessity for huge amounts of training data low at the same time. We further capture the type of the current activity as ACTIVITYTYPE and categorize it as one of four different types ($\in$ {"announcement", "question", "confirmation", "wait"}). Furthermore, the PROMPT parameter acquires the system utterance prompted by the dialog manager prior to recording the user input (Langkilde et al. 1999; Walker et al. 2002). We can proceed on the assumption that the PROMPT in most SDS strongly correlates with the ACTIVITY, since the ACTIVITY can be considered as identifier for the respective prompt. Here, the PROMPT is modeled as entire text string analogous to the UTTERANCE parameter and not as bag-of-words feature vector. The number of words per system turn is captured in WPST and constitutes the counterpart to the previously introduced WPUT (Litman et al. 1999; ITU 2005).

Mostly relevant information about the system–user interaction is the knowledge about whether the current exchange has previously been prompted after an ASR rejection or after a time-out event. A reprompt is the system's repeated attempt to get a valid user response for a predefined slot. We track this event with the parameter REPROMPT? $\in$ {true, false} (Langkilde et al. 1999; Walker et al. 2002) and count the number of reprompts up to the current exchange e_n in #REPROMPT:

$$\#\text{REPROMPT} = \sum_{i=1}^{n} x \begin{cases} 1 & \text{REPROMPT?} = \text{true}, \\ 0 & \text{otherwise}, \end{cases} \tag{3.26}$$

(Langkilde et al. 1999; Walker et al. 2002; Levin and Pieraccini 2006), as well as the number of reprompts within the previous w turns prior to e_n as {#}REPROMPT:

$$\{\#\}\text{REPROMPT} = \sum_{i=n-w}^{n} x \begin{cases} 1 & \text{REPROMPT?} = \text{true}, \\ 0 & \text{otherwise}, \end{cases} \tag{3.27}$$

where w is the size of the window. The average occurrence of reprompts in comparison to slots that require only a single prompting is stored in the parameter %REPROMPT, which is the percentage of reprompts in all previous exchanges:

$$\%\text{REPROMPT} = \frac{1}{n} \sum_{i=1}^{n} x \begin{cases} 1 & \text{REPROMPT?} = \text{true} \\ 0 & \text{otherwise}, \end{cases} \tag{3.28}$$

(Langkilde et al. 1999; Walker et al. 2002). The number of times that have been necessary so far to elicit a desired response from the user for the current activity is captured in the parameter ROLEINDEX. It captures the number of reprompts for a specific activity. Furthermore, we gather the function of the current system turn in the variable ROLENAME $\in$ {"collection", "confirmation", "statement"}. "Collection" refers to a system prompt that requests information from the user, e.g., "Please say the brand name of your modem." or "At what time are you leaving?". "Confirmation" refers to prompts that clarify uncertainties after observing low ASR confidence and demand a yes-no-answer from the user, e.g., "You want to leave at 9 a.m., is that right?". "Statement" refers to prompts that serve as fillers ("Okay, let me check that for you.") or that provide information ("Your next bus leaves at 7 p.m. at Bloomfield.").

All proposed parameters are summarized in Table 3.2.

3.2.2 Emotion Annotation

The issue of annotating a data corpus with emotional states has been discussed in Sect. 2.2.3. In the following, we present the annotation scheme that has been applied when annotating the emotions in the real-life corpora employed in this work. Furthermore, the result of the annotation is discussed. The data corpus employed in the majority of experiments centered around emotion recognition stems from the SpeechCycle Broadband Agent (SC-BAC). Additionally, data from the CMU Let's Go corpora LG-FIELD and LG-LAB have been labeled by a single annotator. Moreover, a third-party corpus from a large German telecommunication company and a Wizard-of-Oz corpus from children interacting with a robot are applied for evaluating some emotion recognition techniques. However, the description of these corpora is postponed to Sect. 4.3.1 since they have not been annotated by ourselves.

Table 3.2 Interaction Parameters modeling a single system-user exchange and serving as the model's input variables, derived from the ASR, SLU, and DM modules

ASR	SLU	DM
ASRConfidence	HelpRequest?	ActivityTrigram
ASRRecognitionStatus	OperatorRequest?	ActivityType
Barge-in?	SemanticParse	Activity
ExMo	#OperatorRequests	DialogDuration
Grammar	#HelpRequests	Loop Name
MeanASRConfidence	%OperatorRequests	Prompt
Modality	{#}HelpRequests	Reprompt?
Triggered Grammar	{#}Operator Requests	RoleIndex
UserTurnDuration		RoleName
UnexpectedModality?		WordsPerSystemTurn
Utterance		#Exchanges
WordsPerUserTurn		#Reprompt
#Barge-ins		#SystemQuestions
#ASRRejections		#SystemTurns
#ASRSuccess		#UserTurns
#TimeoutPrompts		%Reprompt
#Timeouts_ASRRej		{#}Reprompt
#UnexpectedModality		{#}System Questions
%Barge-ins		
%Timeouts_ASRRej		
%ASRSuccess		
%TimeoutPrompts		
%UnexpectedModality		
{Mean}ASRConfidence		
{#}Barge-ins		
{#}ASRRejections		
{#}ASRSuccess		
{#}TimeoutPrompts		
{#}Timeouts_ASRRej		
{#}UnexpectedModality		

As most users show only negative emotions toward real-life SDS, other emotions than Anger are not considered. It should be noted that apart from Anger, other emotions can sometimes be observed, such as Happiness or Boredom. As an increasing number of classes raises the probability of confusions and as their relevance for predicting critical dialog situations is not given, all remaining emotions, including the neutral state, are in the following summarized as NonAnger.

Annotation Scheme

SC-BAC comprises 22,774 user utterances from 1,911 individual users that interacted with the system by using spoken commands. Three annotators labeled the data using the annotation capability of Witchcraft, a tool to support research in online monitoring (see Sect. 3.3). The following labels and instructions have been provided:

- NONANGER: the user does not display anger of any type. This includes neutral utterances and emotions other than anger.
- SLIGHTANGER: the user displays (slightly) negative emotions, i.e., the user seems on the edge, sounds bad-tempered, ill-humored, dour, stressed out, or slightly pushy.
- STRONGANGER: the user displays acoustically strong negative emotions. Anger is clearly noticed, and the speech can contain users that are furious, mad, and shout at the agent.
- GARBAGE: incomprehensible utterances, such as TV noise, mumble, cross-talk, off-talk and non-speech events, i.e., a baby crying in the background, barking dogs, street noise, etc.

Due to the shortness of most samples, each utterance in all corpora was annotated with one single label since the complete spoken entity did only show one predominating emotional state. The annotation scheme depicts a tradeoff between a too large number of classes, leaving a rather free choice to the annotator but complicating the selection, and a binary scheme, forcing the annotator into false decisions.

Random sampling turned out that emotions other than negative and neutral ones, such as boredom, happiness, sadness, etc., were virtually not present in the data. Neutral speech and other emotions are thus categorized as NONANGER. All samples from the database were presented to the annotators in chronological order and independent of the other two annotators. In this way, the history of a turn was known and the content of an utterance was intentionally preserved. The annotators, two advanced students of engineering and one author of this book, were familiar with the respective voice portal and linguistic emotion theory. Prior training was intentionally omitted in order to leave room for interpretation.

Inter-Rater Agreement

During the annotation it became evident that deciding if an utterance was to be labeled as SLIGHTANGER or NONANGER often proved very difficult, which was also the case with utterances fluctuating somewhere between SLIGHTANGER and STRONGANGER. Additionally, many utterances contained a mixture of speech from interactions with non-involved persons ranging from normal chattering to heated arguments. Some users switched the telephone to hands-free mode and conversed with the system from a distant place. This led to a complication of the annotation task. The difficulties the annotators faced becomes evident when calculating the κ-value based on the labeled data. Cohen's Kappa has been discussed Sect. 2.1.2. In Table 3.3 one would assume that the three annotators agreed on a very large proportion of the utterances. A mean κ of 0.70 can be interpreted as substantial agreement of the three annotators (cf. Landis and Koch (1977) or Sect. 2.1.2). It can be further noted that this value ranges among the highest reported in other studies (see Sect. 2.2.3).

However, focusing on the κ scores from the four possible labels it becomes evident that each category STRONGANGER, SLIGHTANGER, NONANGER, and GARBAGE

Table 3.3 κ values measuring agreement on the entire corpus

Annotator	1 & 2	1 & 3	2 & 3	Mean
p_A	0.96	0.96	0.93	
p_R	0.81	0.85	0.83	
κ	0.80	0.71	0.59	0.70/68,133 pairs

p_A depicts agreement on all samples in percent, p_R the expected *random* agreement

Table 3.4 κ values obtained from utterances labeled as NONANGER by at least one of the compared annotators

Annotator	1 & 2	1 & 3	2 & 3	Mean
p_A	0.97	0.96	0.93	
p_R	0.90	0.95	0.93	
κ	0.66	0.23	0.05	0.31/64,386 pairs

p_A depicts the agreement on all samples in percent, p_R the expected *random* agreement

Table 3.5 κ values obtained from utterances labeled as GARBAGE by at least one of the two compared annotators

Annotator	1 & 2	1 & 3	2 & 3	Mean
p_A	0.85	0.74	0.64	
p_R	0.85	0.62	0.64	
κ	0.03	0.33	0.00	0.12/4,050 pairs

p_A depicts the agreement on all samples in percent, p_R the expected *random* agreement

Table 3.6 κ values obtained from utterances labeled as SLIGHTANGER by at least one of the compared annotators

Annotator	1 & 2	1 & 3	2 & 3	Mean
p_A	0.47	0.46	0.17	
p_R	0.51	0.39	0.32	
κ	−0.08	0.13	−0.22	−0.06/3, 963 pairs

p_A depicts the agreement on all samples in percent, p_R the expected *random* agreement

reaches very differential agreement. In Table 3.4 the average κ can be reported with 0.30. Since the corpus mainly consists of utterances labeled as NONANGER the agreement by chance is rather high. By the nature of the κ coefficient the resulting value is very small due to the high a priori probability for the NONANGER class.

Utterances labeled as GARBAGE are by nature very differential, which has an impact on the agreement values from Table 3.5. The mean agreement from all three annotators on which utterance is considered as GARBAGE yields 74.3 % (p_A). Since the samples include recordings where speakers think aloud, talk to a third person, or slur in an almost incomprehensible way they have been sometimes categorized as speech and in other cases as GARBAGE turns, explaining the rather low agreement.

In Tables 3.6 and 3.7 the negative κ values indicate that Annotator 2 has a different conception of SLIGHTANGER and STRONGANGER than Annotators 1 and 3, who in

Table 3.7 κ values obtained from utterances labeled as STRONGANGER by at least one of the compared annotators

Annotator	1 & 2	1 & 3	2 & 3	Mean
p_A	0.38	0.47	0.46	
p_R	0.40	0.42	0.46	
κ	−0.04	0.07	−0.01	0.01/993 pairs

p_A depicts the agreement on all samples in percent, p_R the expected *random* agreement

Table 3.8 κ values over utterances labeled as ANGER (STRONGANGER and SLIGHTANGER collapsed) by at least one of the compared annotators agreement

Annotator	1 & 2	1 & 3	2 & 3	Mean
p_A	0.59	0.58	0.29	
p_R	0.60	0.45	0.43	
κ	0.04	0.23	−0.24	−0.02/4, 320 pairs

p_A depicts the agreement on all samples in percent, p_R the expected *random*

turn do agree slightly. Particularly, the agreement of 17 % between Annotator 2 and 3 for the label SLIGHTANGER is unexpected. A rating by chance would have led to higher results here. At this point cultural differences in the perception of negative emotions may be a possible reason for this outcome. Annotators 1 and 3 were Europeans, annotator 2 was Asian. Another possible explanation would be the accidental wrong choice of the label.

When mapping the STRONGANGER and SLIGHTANGER labels to one label category (Table 3.8) the κ and agreement values indicate the complexity faced by the classifier to separate the data in the later experiments.

Gender Annotation

In a second step, the gender of the speakers was manually annotated. Each annotator listened to all 1,911 dialogs, However, only to the first utterance of each dialog. The annotator determined the gender class, which was one of MALE, FEMALE and UNSURE. In case the gender could not be reliably determined by the annotator, he could request the second, third, etc. utterance. Once the annotator felt sure about his decision, he could automatically assign the gender label to all remaining utterances of this speaker which eventually sped up the rating process significantly. The final label was determined based on majority voting, where the turns labeled as GARBAGE from the anger annotation tasks have been automatically assigned to UNKNOWN. Further, the utterances of eight speakers were assigned to UNSURE since all three annotators differed. In total the corpus contains 10,649 female, 10,844 male und 1,281 unknown utterances, which results in 942 female, 915 male, and 54 unknown dialogs, see Table 3.9.

Table 3.9 κ values for annotation the speakers' gender (male, female, unknown)

Annotator	1 & 2	1 & 3	2 & 3	Mean
κ	0.982	0.969	0.960	0.970/1,911 pairs

Table 3.10 Distribution of *anger* in the SpeechCycle Levelone Broadband Agent corpus

Class	Quantity	Percentage (%)	# Speakers	Avg. # of Exch./Dialog
All utterances				
NONANGER	20,500	90.3	1,872	13.3
SLIGHTANGER	770	3.4	393	13.4
STRONGANGER	161	0.7	92	10.8
GARBAGE	1,167	5.2	572	17.9
Undefined	113	0.5	90	13.3
Female speakers				
NONANGER	10,102	94.9	962	13.2
SLIGHTANGER	402	3.8	215	12.9
STRONGANGER	70	0.7	40	8.7
Undefined	75	0.7	53	12.6
Male speakers				
NONANGER	10,346	95.4	925	13.4
SLIGHTANGER	368	3.4	178	13.9
STRONGANGER	91	0.8	52	12.3
Undefined	38	0.4	37	14.6

Gender annotations yield a much higher agreement, as can be seen from the obtained κ value of 0.97. Classifying MALE versus FEMALE might thus achieve much higher performance scores in contrast to an ANGER versus NONANGER task, since the pattern frontiers seem to be less blurred in the gender classification task. The gender annotation for the two Let's Go corpora was conducted based on the same procedure.

Analysis

The overall distribution of anger in the corpus is depicted in Table 3.10. While a data or speech corpus of 22,711 non acted audio files is rather large, the actual proportion of emotional speech in this corpus is relatively small.

3.35 % turns contain SLIGHTANGER and 0.72 % contain hot STRONGANGER speech. The number of individual callers showing emotions is even lower, approximately by a factor of 0.5. The undefined utterances are the ones all three annotators could not agree on a class. According to the statistics a user showing SLIGHTANGER or STRONGANGER displays a negative emotion twice per dialog. Referring to the average dialog length, it is obvious that dialogs which the user is not pleased with (and therefore angry) do not last as long as other dialogs.

Regarding the as GARBAGE labeled utterances, there are 1,167 utterances originating from 572 individual speakers. Thus, approximately 19 % of the speakers produce a GARBAGE utterance twice per dialog. The GARBAGE utterances per speaker ratio is similar to the ratios of utterances annotated as SLIGHTANGER or STRONGANGER per speaker. Also, dialogs containing GARBAGE utterances last longer. This is probably due to the cause that the users needed to repeat their sentences several times. Evidently, both genders are equally represented in this corpus. While female speakers tend to end the call earlier when SLIGHTANGER or STRONGANGER, male speakers are not inclined to end the call that early. Women showed a higher amount of SLIGHTANGER utterances compared to male speakers, whereas men showed a higher amount of STRONGANGER utterances compared to female speakers. This tendency, however, turned out not to be statistically significant. The comparison of women and men further shows the lower rate of undefined labels among the male speakers. It seems that the annotators that were all male had less difficulties to agree on a label with the male speakers than with the female speakers.

Let's Go Field and Let's Go Lab

For comparison reasons the two Let's Go corpora LG-FIELD and LG-LAB have been annotated with negative emotional states. The character of the Let's Go field corpus (LG-FIELD) differs from the SpeechCycle corpus (SC-BAC), although both data sets originate from real-life systems. Most users from the Let's Go system use mobile phones to connect to the system and call while on the move directly from the street or pub where the background noise is even stronger than in the SpeechCycle data. Many recordings are distorted. The Let's Go Lab corpus annotations give an indication whether users behave differently in laboratory conditions compared to real-life scenarios.

According to the lessons learned from the previous annotation process, the Let's Go corpora have been annotated with a more fine-grained, continuous emotion scheme, mirroring the degree of anger as in Burkhardt et al. (2006). This allowed a more detailed description of the negative emotional state. The labels are as follows:

- *1: friendly*, incl. emotions such as happiness,
- *2: neutral*, incl. other emotions, such as boredom, sadness, etc.,
- *3: slightly angry*, the user displays slightly negative emotions, i.e., the user seems on the edge, sounds bad-tempered, ill-humoured, dour, stressed out or slightly pushy,
- *4: angry*, anger can be clearly noticed,
- *5: very angry*, the user displays acoustically strong negative emotions, i.e., clear rage, hot anger, fury, madness,

Table 3.11 Distribution of *anger* in the Let's Go Field corpus. Female speakers show a significantly higher amount of SLIGHTANGER on this data set ($p < 0.05$)

Class	Quantity	Percentage (%)	# Speakers	Avg. # of Exch./Dialog
All utterances				
NONANGER	3,309	68.5	289	27.8
SLIGHTANGER	693	14.3	134	34.9
STRONGANGER	241	5.0	31	44.3
GARBAGE	589	12.2	184	31.63
Female speakers				
NONANGER	1,857	68.3	962	28.8
SLIGHTANGER	488	17.9[a]	215	36.6
STRONGANGER	107	3.9	40	46.9
GARBAGE	268	9.8	53	34.4
Male speakers				
NONANGER	1,383	71.2	925	26.5
SLIGHTANGER	191	9.8[a]	178	32.4
STRONGANGER	134	6.8	52	39.1
GARBAGE	235	12.1	37	32.1

- *g: garbage*, non-speech events, such as coughing, sneezing, dogs barking, mumble, baby crying in the background, strong street noise, etc.

The corpora comprising 4,832 user turns (LG-FIELD) and 2,036 (LG-LAB) were annotated by a single annotator only, who was a German female advanced student of computer science and who was not involved in the SC-BAC annotation. She was trained with reference samples from NONANGER, SLIGHTANGER and STRONGANGER utterances from SC-BAC and could fall back to the samples during rating. To assure the validity of the ratings, a second annotator who was also involved in the SC-BAC rating process rated a subset of 10 % randomly selected calls. Likewise the SC-BAC annotation scheme, a chronological order was preserved, i.e., the annotators listened to the dialog in chronological and not random order. The agreement between the two annotators on the subset can be reported with *Cohen's* $\kappa = 0.61$ (LG-FIELD) and $\kappa = 0.65$ (LG-LAB) which is, according to Landis and Koch (1977) still substantial agreement. Spearman's ρ can be reported with 0.76 and 0.81, respectively. However, as final label for both corpora the labels of the single annotator apply. To obtain comparability with SC-BAC, the new scheme for LG-FIELD and LG-LAB was mapped to the SC-BAC scheme, i.e., the classes "5" and "4" were collapsed to STRONGANGER, "3" was attributed SLIGHTANGER, "2" and "1" were collapsed to NONANGER. Comparability between the two annotation schemes and the respective corpora remains since the collapsed classes share the same properties. Results are depicted in Tables 3.11 and 3.12.

As we can see, the amount of user anger in the LG-FIELD corpus is considerably higher with 14.3 % SLIGHTANGER and 5.0 % *StrongAnger* utterances compared to SC-BAC. For comparison, the backup-annotator rated 17.3 % SLIGHTANGER and

Table 3.12 Distribution of *anger* in the Let's Go Lab corpus

Class	Quantity	Percentage (%)	# Speakers	Avg. # of Exch./Dialog
All utterances				
NONANGER	1,552	76.2	140	26.4
SLIGHTANGER	337	16.6	64	32.8
STRONGANGER	26	1.3	8	33.0
GARBAGE	121	5.9	58	27.5
Female speakers				
NONANGER	775	78.7	77	26.6
SLIGHTANGER	185	18.8	36	34.7
STRONGANGER	25	2.5	7	32.0
GARBAGE	73	7.4	34	26.8
Male speakers				
NONANGER	770	83.5	62	26.3
SLIGHTANGER	151	16.3	27	31.4
STRONGANGER	1	0.1	1	33.1
GARBAGE	48	5.2	24	28.0

10.5 % STRONGANGER turns on the subset. Female speakers show a significantly higher amount of SLIGHTANGER on this data set ($p < 0.05$). According to the annotator the LG-LAB corpus contains 16.6 % SLIGHTANGER and 1.3 % STRONGANGER utterances. This is basically similar to LG-FIELD, except that the users in the laboratory virtually did not show STRONGANGER, which might be attributed to the presence of the investigator. Here the backup-annotator rated 15.4 % SLIGHTANGER and 1.6 % STRONGANGER turns on the subset.

3.3 A Workbench for Supporting the Development of Statistical Models for Online Monitoring

The use of online monitoring may enhance the basic architecture of an SDS in that it delivers additional knowledge about the user and the interaction. The information may, e.g., be derived from data gathered in the course of the interaction, such as interaction log data and audio recordings, which is subject to statistical classifiers. As we have previously seen, this allows adaptivity in SDS based on a large number of predicted target variables, such as age, gender, specific speakers, emotional states, user satisfaction, task success probability, etc. Currently, such statistical models are rarely deployed in dialog systems. In fact, novel techniques developed in research studies are to a large extent merely subject to batch evaluation, which is usually performed in cross-validated or fixed splits, where, e.g., 70 % of the data are employed for training and the remaining 30 % for testing. For novel data-driven techniques targeting on online deployment the question has to be raised, which impact the deployment of a specific model would have on a dialog when being employed in a live system. Where in a dialog would it fail and where would it deliver acceptable predictions? It seems to be important to emphasize that at latest when the deployment of such models is scheduled their direct influence on *specific* dialogs should be of highest interest to both system developers and research scientists. Going live and shifting a classifier on the dialog level may cause critical situations and may have a severe and also negative impact on the dialog. Simply imagine a dialog system endowed with new "intelligence" would address a female speaker as "Sir", tries to calm down obviously non-angered speakers and—in IVR applications—escalates callers to an agent although the interaction has been running well at the time of escalation. For future evaluation tasks of the addressed prediction models it will become ever more important to consider the context of the interaction jointly with the knowledge of the dialog system. Further questions may demand an answer. We can assume that with increasing available data collected during the interaction, which constitutes an increasing experience with a specific user, a prediction becomes more reliable and accurate. Our future interest may be in answering, at what point in time statistical models predicting gender, speaker age, and expert status would deliver a reliable statement that can indeed be used for adapting the dialog.

3.3.1 Requirements Toward a Software Tool

For the research in this field we felt the need for a software tool that supports the development and evaluation of such novel data-driven techniques. This software needs to address three basic requirements:

- data management and presentation,
- data annotation,
- model evaluation.

Data Management and Presentation

Statistical models presented in this book are based on massive data sets that are usually stored in the form of logged text files, such as in the CMU Let's Go system. The data employed may contain 100,000 dialogs and more per corpus. How system and user conversed is hardly replicable in this representation. For corpora of this size a sophisticated framework is required that allows a fast and random access to the data and that summarizes an interaction between the system and the user in a rapidly ascertainable way. In summary, the tool needs to be able to

- serve as central *front-end* to massive amounts of dialog data, along with annotations and transcriptions, filed on a central storage,
- allow the *management of multiple dialog corpora* at the same time,
- allow the user to *sort, group, store, and comment* on dialogs,
- provide a *search functionality* for identifying relevant dialogs,

- *visualize logged interactions* in a way that makes interaction data readable and that outlines central aspects of a dialog,
- *simulate the logged interaction* by playing back the recorded dialog synchronized with log information,
- allow *multi-user support*, i.e., multiple workspaces and multi-user access to the centrally stored data that enables, e.g., simultaneous data analysis and annotation.

Data Annotation

Target variables in statistical modeling in the SDS domain require usually an annotation with appropriate labels on three annotation levels, depending on the task of the model. The software tool needs to provide functionality for annotating labels

- on *user utterance level*, e.g., for annotating age, gender, etc. based on recorded user utterances,
- on *exchange level*, e.g., for annotating satisfaction scores,
- on *dialog level*, e.g., for allowing an annotation of task success, overall quality etc.

The labels need to be stored in the same central database.

Model Evaluation

Furthermore, the tool requires an extensive capability for evaluating statistical prediction models. Therefore it needs to

- provide an *interface to arbitrary machine learning predictions* from statistical models and recognizers,
- allow the *evaluation of regressive and discriminative classifiers*,
- provide *standard metrics* such as accuracy, recall, precision, etc. for a dialog-wide evaluation,
- *visualize predictions* from classifiers jointly with the dialog flow.

In the dialog system context, a large number of software tools exist, however, their central purpose is the annotation of speech and SDS data. For example, the NITE XML Toolkit (NXT) (Bernsen et al. 2002; Carletta et al. 2003, 2005) offers a generic annotation framework that aims to overcome different data formats employed in annotating spoken language-related and linguistic data by proposing an XML-based annotation format. NXT was motivated by the increasing complexity of language corpora as a result of the growing importance of multi-modal applications. To a certain degree it supports the annotation of multi-modal dialog data. Moreover, it features the annotation of texts, monologues and uni-modal dialog data. In addition to data handling, NXT is endowed with a query language and enables building graphical interfaces. Similar to NXT, the MATE workbench (McKelvie et al. 2001) allows XML-based customizable annotation schemes for speech and text. The annotation tool Transcriber[1] (Barras et al. 2001) was designed for transcribing broadcast news and uses annotation graphs to simplify annotations. DialogView[2] (Yang and Heeman 2005) allows dialog annotations, e.g., tagging speech repairs and speech acts. Dialogs

[1] http://trans.sourceforge.net.

[2] http://cslu.cse.ogi.edu/DialogView/.

can be displayed in different views. For example, the word view depicts words time-aligned with the audio signal while the utterance view shows the entire dialog as a play script.

3.3.2 The Workbench

While the presented tools might be a good choice for annotating dialog data, they do not depict a general front-end to SDS corpora and particularly do not support development and evaluation of statistical prediction models for SDS. In the following, we present such a software framework that we developed in the last years. It was perpetually developed further until it reached the degree of maturity to make it publically available.[3] The software framework is called "Workbench for Intelligent exploraTion of Human ComputeR conversaTions" (WITcHCRafT) and we denote it as *workbench*, since it deals with a large number of tasks centered on statistical prediction models for SDS. Moreover, it allows managing, mining and analyzing large dialog corpora. It brings logged conversations back to life such that it simulates the interaction between user and system based on system logs, audio recordings and a TTS engine. Furthermore, it offers annotation views that can be used to assign labels to the dialog data on different levels. Finally, it simulates the deployment of statistical models and allows their evaluation. A screenshot of the Analysis Perspective in Witchcraft is depicted in Fig. 3.6.

3.3.3 Data Management

Witchcraft presents the complete dialog containing system prompts, system actions, ASR accuracy, recognized word strings, parsed semantics, barge-in behavior, etc. in a structured and easily accessible manner at each dialog step, see in Fig. 3.6. The Witchcraft user can jump into any position within the dialog and start replaying. System prompts are synthesized with the text-to-speech (TTS) engine OpenMary (Schröder and Trouvain 2003) in case system prompts and user utterances do not exist. Otherwise the user utterances are played back from original and logged recorded conversations. The *Main View* Ⓕ is thereby temporally synchronized with the *Wav Player View* Ⓗ, which is able to replay entire recordings or a concatenation of user recordings. The latter allows the Witchcraft user to rapidly gain an overview of the acoustic behavior of the user. Relevant dialog-wide details are depicted in the *Call Detail View* Ⓒ, such as the average ASR confidence, the number of exchanges, the duration etc. The dialog can be further displayed as play script in the *Conversation View* Ⓖ.

[3] http://sourceforge.net/projects/witchcraftwb.

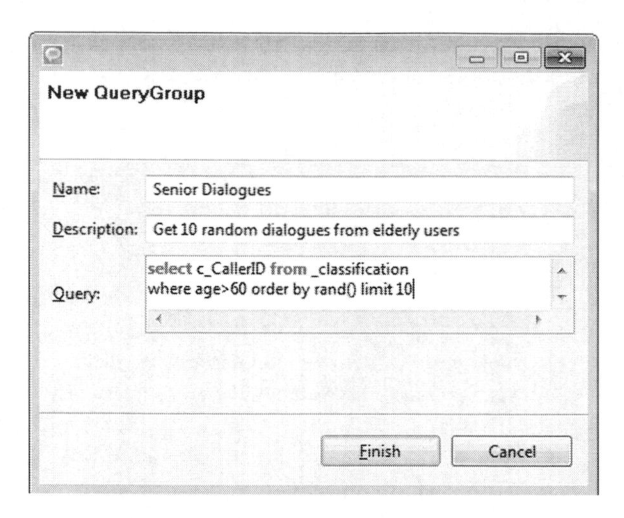

Fig. 3.4 Creating a QueryGroup in Witchcraft

In data sets with several ten thousand dialogs, structural storage principles may be of valuable help. Witchcraft features a grouping functionality within the *Dialog Storage View* Ⓑ, allowing to group dialogs by their unique IDs into *IDGroups* or by an SQL[4] database query into *QueryGroups*. Examples for such QueryGroups are dialogs from male users, dialogs with specific content, e.g., users who lost their password or users requesting the bus with number "26X", dialogs lasting at most 10 min, etc. An example of creating a QueryGroup in Witchcraft is depicted in Fig. 3.4. Interesting or relevant dialogs discovered in the examination process can be stored in *IDGroups*, which allows, e.g., to store dialogs containing angry users in one group identified by their IDs. To load dialogs from the database, either an SQL query may be entered in the search dialog or IDGroups and QueryGroups can be dragged and dropped into the *Workspace* Ⓐ. There, all dialogs matching the search query or which are contained in an IDGroup or QueryGroup are listed and may be opened for further analysis or annotation.

The architecture of Witchcraft allows access to an unlimited number of dialog corpora (see *Corpus View* Ⓓ in Fig. 3.6) that are stored in an SQL database server. An integration of new corpora is straight-forward, but requires some manual steps. Witchcraft demands a new SQL database containing two tables. The *dialogs* table hosts dialog-level information affecting the dialog en bloc, such as the dialog ID, the category, filename of full recording, etc., and the *exchanges* table containing the turn-wise interactions, such as the dialog ID, turn number, system prompt, ASR parse, ASR confidence, semantic interpretation, hand transcription, utterance recording file, barged-in information, etc. Both tables are linked by a 1:n relationship using the DialogID as primary and foreign key, i.e., one entry in the dialogs table relates to n entries in the interactions table, cf. Fig. 3.5.

Corpora-related sound files containing full recordings or single user prompts are stored in the Witchcraft workspace folder.

---

[4] Structured Query Language, see e.g., Molinaro (2005).

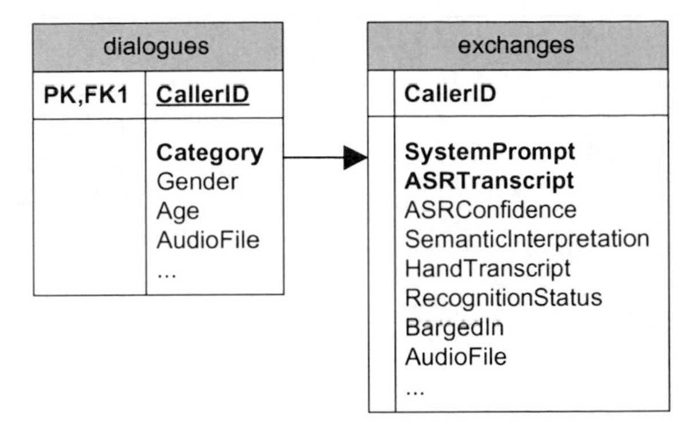

Fig. 3.5 Entity-relationship diagram of dialogs and exchanges table with 1:n relationship. Bold database columns are required, the others are optional

3.3.4 Evaluating Statistical Prediction Models for Online Monitoring

Unlike NXT, the MATE workbench, Transcriber, and DialogView, whose purpose is the annotation of data, Witchcraft is predominantly not an annotation tool. Although Witchcraft can be also used for annotation its central purpose is a different one: The workbench contrasts dialog flows of specific dialogs, which are obtained from a dialog corpus with the estimations of arbitrary prediction and classification models. By that, it is instantly visible which knowledge the dialog system would have at what point in time in the dialog. That means Witchcraft is able to visualize components of the knowledge vector that was proposed in Fig. 1.5 in Chap. 1. Imagine a dialog system would be endowed with an anger recognizer and a recognizer that predicts the outcome of a dialog, i.e., task success. Each of the recognizers would be designed to deliver an estimation at each exchange. The two recognizers report to the dialog manager whether the user is angry and how likely system and user will successfully solve the task, based on what the statistical model has "seen" so far from the dialog. Witchcraft now depicts to what extent the recognizers deliver a correct result. This is illustrated in Ⓘ in Fig. 3.6 in the Model Chart Views called "anger_speaker_independent" and "new task_completion_model" as well as in the Prediction Table Views Ⓚ, tabular representations of the charts that also display misclassifications. Using Witchcraft and appropriate models we can deduce, e.g., when

- the task success prediction model recommends a transfer to a human agent,
- the system "thinks" the user is angry based on the anger classification model,
- the system seems to be certain that it is talking to a male/female or junior/senior person based on gender and age models.

Classifiers supported by Witchcraft can be basically categorized into one of the three categories in supervised learning, see also Sect. 2.1:

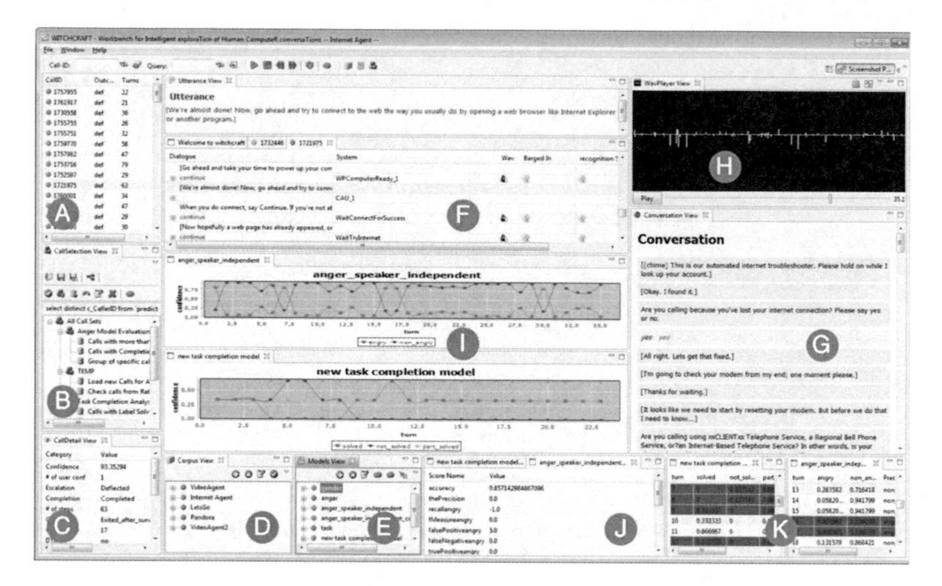

Fig. 3.6 The Witchcraft Workbench in the analysis perspective. **a** *Workspace* hosts a selection of dialogs that match a specific SQL query. **b** *Dialog storage view* allows to group and store dialogs according to specific characteristics. **c** *Dialog detail view* depicts important information about a specific dialog. **d** *Corpus view* allows to switch between different dialog corpora. **e** *Models view* hosts the definition of prediction models that can be applied to the dialogs. **f** *Main view* displays the interaction of the currently opened dialog. **g** *Conversation view* displays the dialog as play script. **h** *WavPlayer view* allows to play back the recorded dialog. **i** *Chart views* display the prediction of the activated models. **j** *Model performance views* display performance values of the activated models when applied on the opened dialog. **k** *Prediction table view* table representations of the chart views

Discriminative Binary Classification Discriminative classifiers are used to predict a distinct, discrete class. A binary classifier may discriminate between two classes, where one class is defined as the positive, the other as the negative one. An example for such a classifier would, e.g., be a gender recognizer that delivers the discrete predictions 'male' or 'female'. Each possible class label in discriminative predictions often comes along with the classifier's confidence score, e.g., 86 % male, 14 % female, see e.g., Bocklet et al. (2008).

Discriminative Multiclass Classification Multiclass classifiers are of the same nature as the binary classifiers except that they discriminate between more than two classes. Positive and negative classes are not defined here. A classifier discriminating between different age classes can be brought in as an example, see e.g., Metze et al. (2008).

Regression A regressive classifier provides predictions of continuous character and delivers approximations to the (numerical) class label. A regression task in the SDS domain could be a cooperativeness score of the user, or the estimated user satisfaction in percent, see e.g., Engelbrecht et al. (2009), Higashinaka et al. (2010b), and Higashinaka et al. (2010a).

Witchcraft thereby does not contain "intelligence" on its own but makes use of and manages the predictions of external recognizers that host the classifier and the statistical model. Within the Witchcraft context it is assumed that a recognizer is implemented either as stand-alone recognizer or with help of a machine learning (ML) framework. We emphasize that Witchcraft itself does neither perform feature extraction nor classification. Both tasks are outsourced to the recognizer. The workbench operates on the exchange level requesting the recognizer to deliver a prediction based on information available at the currently processed dialog turn of a specific dialog. Where and how the recognizer accomplishes this is not part of the architecture.

In a previous version of Witchcraft (Schmitt et al. 2010a) such an ML framework was directly integrated into Witchcraft. That ML framework of our choice was RapidMiner,[5] formerly known as "YALE" (Mierswa et al. 2006). It covers a vast majority of supervised and unsupervised machine learning techniques. The initial plan to interface other ML frameworks and classification tools, such as MatLab, the R framework (Crawley 2007), BoosTexter (Schapire and Singer 2000), Slipper (Cohen and Singer 1999) and HTK (Young 1994) that are frequently used in research turned out to be unrealizable due to heterogeneous programming languages and non-existing interfaces. Instead, a direct integration of RapdidMiner and further ML frameworks was abandoned and a generic XML interface was introduced instead, allowing to interface an arbitrary ML tool or recognizer. An overview of the interplay between Witchcraft and such an external recognizer is depicted in Fig. 3.7.

The recognizer, which in this case is implemented by use of a Machine Learning framework, performs feature retrieval, preprocessing, and classification. It is then required to deliver XML documents (Bray et al. 2004) that fit the model definition in Witchcraft, see Fig. 3.8.

Each XML document represents the prediction of the recognizer for a specific dialog exchange of a single dialog. For discriminative classification tasks, which, e.g., predict gender or emotion classes, it contains the number of the turn that has been classified, the actual class label, and the confidence scores of the classifier for the respective classes, see Listing 3.1.

Listing 3.1 XML describing a discriminative classifier prediction

```
1 <xml>
2     <turn>
3         <number>1</number>
4         <label>anger</label>
5         <prediction>non-anger</prediction>
6         <confidence class="anger">0.08</confidence>
7         <confidence class="no-ang">0.92</confidence
            >
8     </turn>
9 </xml>
```

[5] www.rapid-i.net.

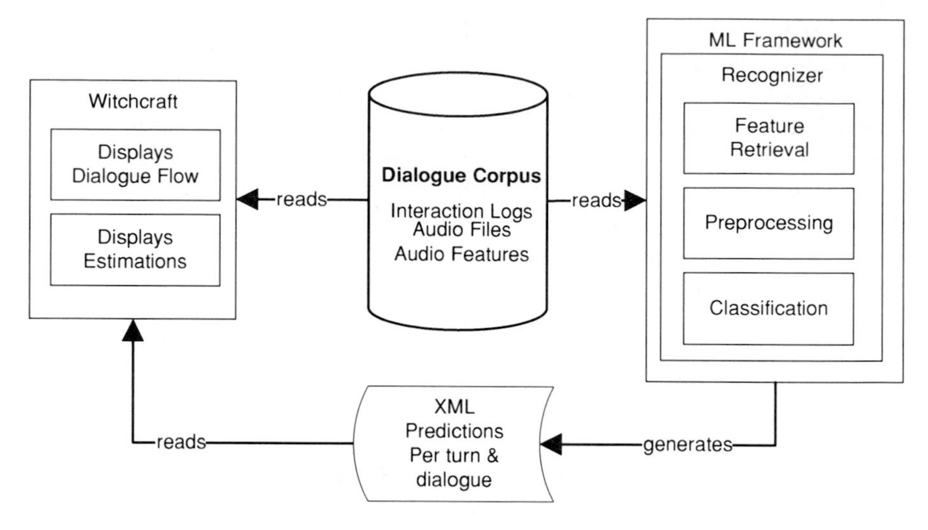

Fig. 3.7 Dependency of Witchcraft and related recognizers implemented within an ML framework

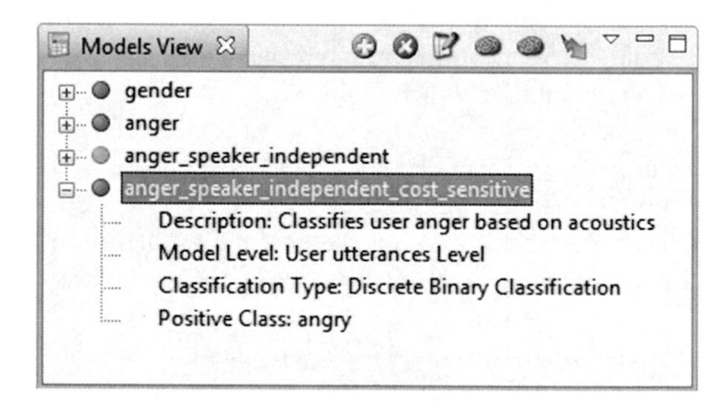

Fig. 3.8 Definition of a model within Witchcraft. External recognizers have to deliver predictions for the defined models as XML documents

In regression tasks, such as the prediction of user satisfaction or estimating cooperativeness scores, the returned XML document needs to contain the turn number, the actual label, and the prediction of the classifier, see Listing 3.2.

Listing 3.2 XML for describing a regressive classifier prediction

```
1  <xml>
2      <turn>
3          <number>1</number>
4          <label>5</label>
5          <prediction>3.4</prediction>
6      </turn>
7  </xml>
```

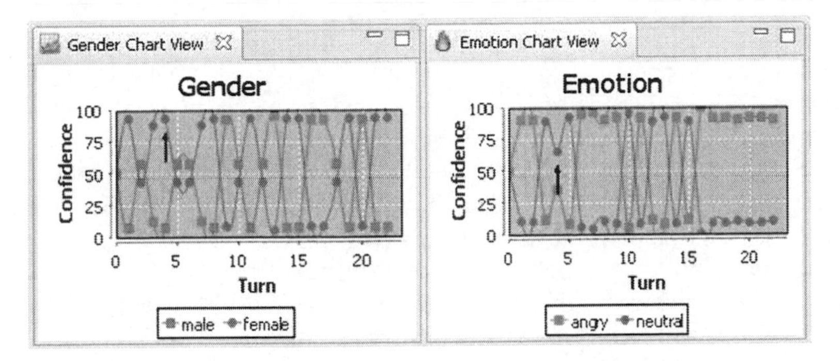

Fig. 3.9 Screenshot of charts in Witchcraft based on turn-wise predictions of an anger and a gender classifier. The *line* with the quadrates in the *left-hand* gender model represents the confidence of the recognizer of observing a male speaker (*line with circles* = female speaker). In the emotion model on the *right* the *line* with the quadrates symbolizes the confidence of the speaker being angry in the current turn (*line with circles* = neutral)

Witchcraft parses the XML predictions into an SQL table, displays the estimations of the recognizer in a chart view, see Ⓘ in Fig. 3.6, and calculates dialog-wide performance scores, such as *accuracy, f-score, precision and recall values, root mean squared error etc.* The metrices are displayed in the *Model Performance View*, see Ⓙ in Fig. 3.6, and give information above how precisely the classifier worked on the dialog level. The classifier's estimates are further depicted in the *Prediction Table View*, see Ⓚ, which allows to spot misclassified exchanges.

After parsing the predictions, Witchcraft allows to search for dialogs with low overall prediction accuracy, dialogs with high true positive rates, low class-wise f-scores, and the like by using an SQL query. Hereby a detailed analysis of the recognizer's performance on the dialog level and possible reasons for the failure can be spotted.

In the following, an example is given that illustrates the use of recognizers in Witchcraft. In Fig. 3.9 we see prediction series of two binary discriminative classifiers that have been applied on a specific dialog of SC-BAC: a gender classifier that predicts the gender based on single user utterances from the current exchange and an emotion classifier that predicts the user's emotional state at the current exchange. It discriminates between the classes ANGER and NONANGER. The classifiers are built as two independent RapidMiner processes that use acoustic features from the respective utterances retrieved from the database to estimate the classes. This process is designed in analogy to Fig. 3.7. The RapidMiner process is further implemented to output predictions in XML format that are parsed by Witchcraft.

Both lines respectively represent the classifier's confidence from 0 to 100 % for each of the two possible classes. In the emotion model the line with circles constitutes the confidence for a NONANGER utterance, the line with the quadrates for an ANGER utterance. Exemplary for the two models we take a closer look at the gender model. It predicts the gender on turn basis, i.e., it takes the current speech sample and delivers estimations on the speaker's gender. As we can see, there are a number of misrecognitions in this dialog, which originates from a female speaker. The recognizer frequently pretends to observe a male speaker, which is obviously not the case. The dialog could be spotted by searching within Witchcraft for dialogs that yield a low accuracy for gender. It turned out that the mistaken turns originate from the fact that the user performed off-talk with other persons in the background. This caused the misrecognition. This finding suggests an improvement for the model design: instead of training the gender recognizer only with female and male speech samples, an improvement might be achieved by using non-speech and cross-talk samples in order to broaden the recognition from two (male, female) to three (male, female, non-speech) classes. Further it appears sensitive, to create a recognizer that would base its recognition on several speech samples instead of only one, as it is the case in this example. In this case the growing availability of speech-data in a progressed dialog would contribute to deliver a more robust result for determining the gender.

3.4 Summary and Discussion

Data-driven statistical modeling to perform online monitoring requires large amounts of data that need parameterization and annotation. In the beginning of this chapter, in Sect. 3.1, we have introduced four real-life data sets from deployed SDS that will be used to evaluate the novel techniques developed in Chaps. 4, 5 and 6.

Parameterization and Annotation
In Sect. 3.2 we have prepared the corpora for machine learning by defining input and target variables. An extensive interaction parameter set for modeling ongoing system-user interactions has been presented in Sect. 3.2.1, which describes the interaction at a specific dialog step with 52 interaction parameters. It will serve as basis for many statistical models introduced at a later stage of this book. The interaction of an exchange is thereby modeled with relevant parameters from the ASR, LU, and DM modules that affect the exchange itself. Furthermore, the modeling unit spans a window prior to the current exchange and covers important system events that occurred in the immediate context. Finally, the overall interaction is mirrored in dialog-wide parameters that happened prior to the current exchange. The parameters can be determined without manual intervention and mirror both, the interaction itself and implicitly also the system design. As a result of the automatic generation of the set, features requiring manual intervention, such as the hand transcription of the user utterance or dialog acts, are intentionally not included in the proposed parameter set. Although it could be shown that the hand transcription gives an improvement over ASR transcripts, e.g., in predicting task success (Langkilde et al. 1999; Walker et al. 2002), we do not consider its deployment as helpful. ASR under noisy conditions, such as in IVR systems, will remain error-prone implying that for any realistic assessment of the model's performance under realistic conditions we have to deal with this ASR error.

In Sect. 3.2.2 we have described the annotation schemes employed in the following emotion recognition studies. It became evident by analyzing the ratings on SC-BAC from the different annotators that annotating emotional states is a highly subjective task. Low agreement scores could be reported, particularly for SLIGHTANGER and STRONGANGER as well as GARBAGE. The manual annotation process resulted in one of the largest English "real-life" corpora for negative emotions in the field. The κ statistic shows that the number of three annotators for an emotion labeling task is paying off at the end. In order to vote out potential erroneous labels it seems sensitive to determine the final label by a majority vote over the three labels. Thus the rather controversial labeling results of SLIGHTANGER and STRONGANGER can to a certain extent be objectified. Two aspects have to be addressed in our discussion: the agreement and the annotation scheme. The overall agreement of 0.7 on the entire corpus constitutes one of the highest reported on anger data. However, it turned out that a broad agreement on which utterances sound angry and which don't could not be reached, which is mirrored by the low kappa values for the negative emotion classes. The findings suggest that potentially a higher consensus could be reached would the experts only annotate independently a small sample. In a subsequent discussion the annotators could analyze their ratings and in particular the obvious cases of disagreement. These cases could then be used to update the annotation instructions, which would form the foundation of the residual rating process. Nevertheless it should be noted that rating emotions is a highly subjective process and the individual opinion on what is considered as anger can be seen as more important than a high Kappa value. By that, we mirror the diversity of the task. Joint discussions between annotators further require a more or less temporal synchronous rating process, a fact that is often hardly viable in practice. To obtain a clear separation of the patterns it seems sensitive to smooth out inconsistent labels by majority voting, as it was the case here. Moreover, the question has to be risen if the segmentation of the user's negative emotional state into two distinct classes SLIGHTANGER and STRONGANGER already at annotation level is appropriate and preferable to a continuous scale as in Burkhardt et al. (2006). Both schemes have their strengths and drawbacks. The lessons learned show that the annotators often had difficulties to decide for one of the classes, which could have been facilitated by the introduction of a larger continuous scale, where the "false" choice of a neighboring label would be less severe.

In summary the anger distribution of the considered corpora is depicted in Fig. 3.10. All data sets contain less anger than initially anticipated. For SC-BAC STRONGANGER can be reported with 0.72 %, for LG-FIELD with 5.0 % and for LG-LAB with 1.3 %. SLIGHTANGER occured in 3.35 % of all utterances in SC-BAC, 14.3 % in LG-FIELD and even 16.6 % in LG-LAB.

Particularly, SC-BAC contains an extremely slight proportion of STRONGANGER. In general, slight anger occurs more frequently throughout all data sets. It is interesting to note that the anger proportion of the Let's Go systems is substantially higher, which might be correlated with the higher amount of garbage turns in these systems. Users provoking ASR errors by strong background noise or non-speech events seem to get annoyed from the re-prompts that are generated in return for mismatched ASR parses based on garbage.

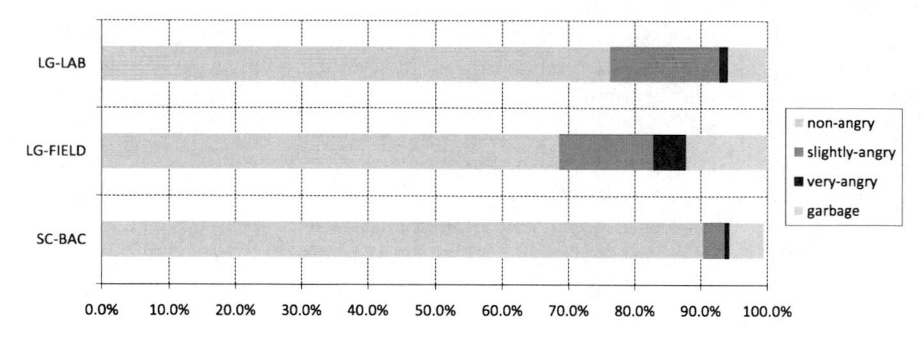

Fig. 3.10 Anger distribution in all three annotated corpora SC-BAC, LG-FIELD, and LG-LAB

The very skew distribution of anger versus non-anger in the addressed corpora will pose a challenge in the classification task and comes close to looking for a needle in a haystack.

Platform Development
For the development of novel strategies for SDS we further presented a software platform that supports the development and evaluation of statistical prediction models, see Sect. 3.3. The Witchcraft Workbench makes alive logged dialogs and supports the data mining process for SDS development. It furthermore visualizes predictions originating from arbitrary classifiers jointly with the dialog flow of a specific dialog. Hereby it depicts the new "intelligence" an advanced adaptive dialog system may achieve when statistical models are deployed for online monitoring. An evaluation of the "detection" phase is thereby rendered possible.

Witchcraft turned out to be a valuable workbench in the everyday work when dealing with large dialog corpora. Several students have been working with this workbench in a multi-user mode to listen, analyze, and annotate dialogs from three different corpora consisting of up to 100,000 dialogs each. Witchcraft allows them to search for dialogs relevant to the current task and provides a comfortable way for obtaining an overview of new corpora. The SQL-based access allows a powerful and standardized query and retrieval of dialogs from the database. Witchcraft offers convenient browsing functionalities and provides central information about a dialog at one glance. Further, it allows to sort and group dialogs for further research and analysis.

The central purpose of the workbench, i.e., the model evaluation functionality, particularly helped to assess the reliability and practical suitability of several statistical models developed in this book, such as the emotion recognizer, see Chap. 4, the Interaction Quality, and user satisfaction classifiers in Chap. 5 and the task success classifier, see Chap. 6. In general, the software platform is intended to provide a basis for a large number of tasks centered on SDS development and analysis. New functionality, e.g., new annotation tasks, can be implemented as plug-in that integrates into the Witchcraft platform. Although it is not the very own intention of Witchraft to perform model-driven evaluation of SDS, it could be used to pinpoint poor dialog design and critical slots based on the predictions of statistical models. Slots where users frequently get upset, or slots where users yield low satisfaction scores or a poor Interaction Quality could be identified by using the search functionality in Witchcraft. An extendibility of the platform is facilitated by using common programming concepts and tools:

- Java: the workbench is implemented in Java, allowing platform independency (Flanagan 1996),
- Plug-in architecture: Witchcraft is based on the rich-client platform (RCP) (Gruber et al. 2005; McAffer and Lemieux 2005; McAffer et al. 2010) of the Eclipse IDE and uses existing, common software and GUI paradigms that come with Eclipse. The advantage of doing so is that all capabilities of the wide spread Eclipse RCP can be used, such as views, perspectives, update, and help functionalities, etc. This again allows a rapid adaptation to new annotation and analysis tasks.
- SQL database: Witchcraft interfaces any JDBC[6]-compliant SQL database server and is deployed with the MySQL database server (Dyer 2005), which is widely in use.

A current limitation of Witchcraft is the lack of an import functionality that allows a convenient data transfer for existing text-based corpora. Instead, the data need to be imported from text-based corpora in semi-manual steps to meet the Witchcraft-compliant table structure, i.e., the column names have to be mapped to the nomenclature of Witchcraft. The various labeling views, allowing the user to annotate emotional states, age, gender, etc. as well as satisfaction scores are currently hard-coded and do not allow a refinement of the labels on the GUI level. Future versions of Witchcraft would need to address this issue.

Witchcraft is an open-source project and thus freely and publically available. It is hosted under GNU General Public License (Smith 2007) at Sourceforge under witchcraftwb.sourceforge.org. The employed component architecture allows for the development of third-party plug-ins and components for Witchcraft without the need for getting into details of the existing code. This facilitates the extension of the workbench by other developers. We included the CMU Let's Go bus information system from 2006 as demo corpus (Raux et al. 2006). It contains 328 dialogs including full recordings. The Witchcraft project includes a parser that allows transforming raw log data automatically from the Let's Go system into the Witchcraft table structure.

[6] Java Database Connectiviy.

Chapter 4
Novel Strategies for Emotion Recognition

In "The Media Equation: How People Treat Computers [...] Like Real People and Places", Reeves and Nass (1996) report on numerous psychological studies showing that users treat computers, i.e., SDS, as real and equivalent interaction partners. Thereby, the interaction behavior differs e.g., when machines are endowed with female instead of male voices. The necessity for a machine to understand the feelings and emotions of its counterpart is therefore a central aspect of "affective computing" (Picard 1997) and may eminently contribute to adaptivity. In field scenarios, it can likewise be observed that users show visible emotions toward SDS. Particularly negative emotions, i.e., user anger, can be noticed when the system makes a mistake or when users discover that they are served by a machine (in telephone-based SDS scenarios). The capability of recognizing the emotional state of the user is a central aspect for online monitoring SDS. This notably holds true for the recognition of critical dialog situations.

When comparing existing studies on emotion recognition, one has to be aware of the precise conditions of the underlying database design, as many of the results published hitherto are based on acted speech data. Some of these databases include sets of prearranged sentences. Recordings are usually done in studios, minimizing background noise, recording speakers (one at a time) multiple times until a desired degree of expression is reached. Real life speech does not have any of these settings. Unlike emotion recognition approaches for laboratory and conversational speech that may take use of larger speech samples the recognition of emotions *in SDS field scenarios* depicts a challenge that has not been sufficiently addressed. Due to the limitations of SDS technology, the user input is frequently restricted to very short sentences or even commands. In return, this poses a challenge to emotion recognition, namely the reliable detection of emotions from limited, noisy, and (with respect to the emotional content) ambiguous speech data.

There are many ways, in which emotional states of humans can be conveyed. However, in speech-based scenarios that aim to recognize the emotional state based on single utterances, two factors prevail: the choice of words, i.e., the linguistic content, and the acoustic variation, i.e., the paralinguistic coloring. When a speaker expresses

A. Schmitt and W. Minker, *Towards Adaptive Spoken Dialog Systems*,
DOI: 10.1007/978-1-4614-4593-7_4,

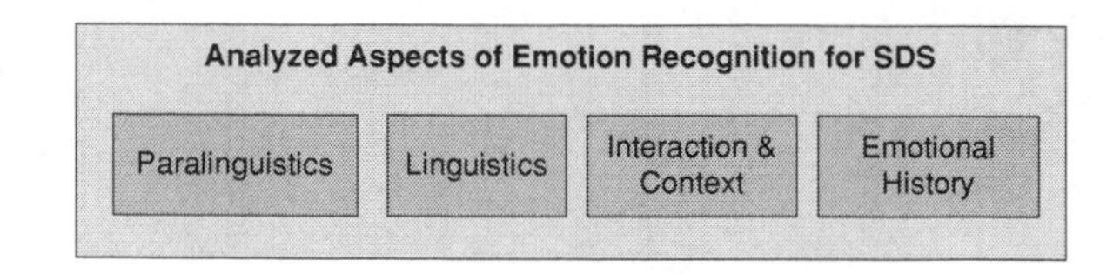

Fig. 4.1 Aspects of speech-based emotion recognition for SDS

an emotion while adhering to an inconspicuous intonation pattern, human listeners can nevertheless perceive the emotional information through the lexical content. On the other hand, words that are not generally emotionally salient can certainly be pronounced in a way, which allows a deduction on the speaker's emotion in addition to the mere lexical meaning. Consequently, our task is first of all to capture the diverse acoustic and linguistic cues that are present in the speech signal and to analyze their correlation to the speaker's emotion. While related work on speech-based emotion recognition has mainly focused on mere paralinguistic modeling (Burkhardt et al. 2008; Kamel and Karray 2007; Morrison et al. 2007) or paralinguistic plus linguistic modeling (Pittermann and Schmitt 2008; Devillers and Vidrascu 2006; Lee and Narayanan 2005), further information that may contribute to a reliable prediction of the user's emotional state particularly adhering in SDS scenarios has only sparsely been explored. In the following, we analyze the dialog context and derive novel dialog-related knowledge sources that may contribute to the distinction of the user's emotional state.

In our approach, we analyze four different aspects related to emotion recognition in SDS, cf. Fig. 4.1.

First, we examine approaches that aim to recognize the user's emotional state from *speech*, i.e., the speech signal and the spoken content in Sect. 4.1. We consider paralinguistic information ("Paralinguistics") and use it for modeling emotions in Sect. 4.1.1. Initially, a common approach is presented that is used in similar form in related work. We will extend this paralinguistic modeling by a more comprehensive and detailed description of the speech signal's acoustic properties in order to obtain better results in automatic classification. In the next Sect. 4.1.2, we discuss and demonstrate to what extent linguistic information ("Linguistics") can be used for prediction and apply statistical techniques from information retrieval to linguistically model emotions.

That followed we examine information sources that exceed the frontiers of a single utterance and direct our attention to *dialog-related* information in Sect. 4.2. An analysis in Sect. 4.2.1 explores to what extent knowledge about the interaction itself may be used to estimate the user's emotional state even without any paralinguistic or linguistic information, but instead, by relying on interaction and context features ("interaction and context"). In Sect. 4.2.2, we furthermore introduce a model that exploits the consecutive occurrence of similar emotional states. This is based on the fact that (negative) emotions rarely occur in isolation as users keep their emotional state for a longer period throughout a dialog. This information is used for modeling the history of the user's present emotion ("emotional history").

Responding to the demands of a real-life emotion recognition system where costs of misclassifying emotions are inequally distributed, we furthermore introduce a cost-sensitive classifier that helps to maximize the number of correctly identified relevant emotional states in Sect. 4.2.3.

Pivotal to the evaluation part in Sect. 4.3 is the section that provides a description of the databases serving as training and test sets for the approaches discussed and evaluated in this chapter. The following evaluation starts by establishing a human baseline and subsequently shows the individual performance of each single subcomponent along with the overall performance of the fused components. We thereby identify relevant features, their individual impact on the task of emotion recognition and simulate the deployment of the proposed emotion recognition system in a real-life scenario. The results are summarized and discussed in Sect. 4.4.

4.1 Speech-Based Emotion Recognition

In the following, we explore emotion recognition approaches, which are exclusively based on information derivable from *isolated spoken utterances*. The proposed approaches may be used for the recognition of the user's emotional state in SDS interaction, but also for the recognition of the speaker's emotions in other contexts, e.g., in monologs and human–human dialogs.

4.1.1 Paralinguistic Emotion Recognition

Paralinguistic information, i.e., information about intonation, volume, pitch, and other acoustic or prosodic features of a speech signal, represents the most central source of information to determine the speaker's emotional state in a spoken utterance. This particularly holds true for frustration, which is first and foremost an acoustic sensation. If we take a closer look at the spectrograms of two utterances from a user talking to an Internet troubleshooter, we can clearly see spectral differences (cf. Fig. 4.2). In both cases, the user says "steady on" to indicate that the LED on the modem is on.

Initially, (cf. Fig. 4.2a) the user speaks normally without visible emotions. After a misrecognition on the part of the SDS, the user gets frustrated and shouts at the dialog system, which is depicted in Fig. 4.2b. The second utterance contains more energy, which is reflected by the higher amount of yellow and red areas in the spectrogram. Especially in higher frequency bands, we observe more energy than in the first utterance and it is clearly visible that the user raised the voice. The second utterance contains a pause at about 0.3 s between "steady" and "on". Presumably, the user expects to facilitate recognition by isolating each word. By drawing upon these differences (such as separating words) in combination with the spectral differences

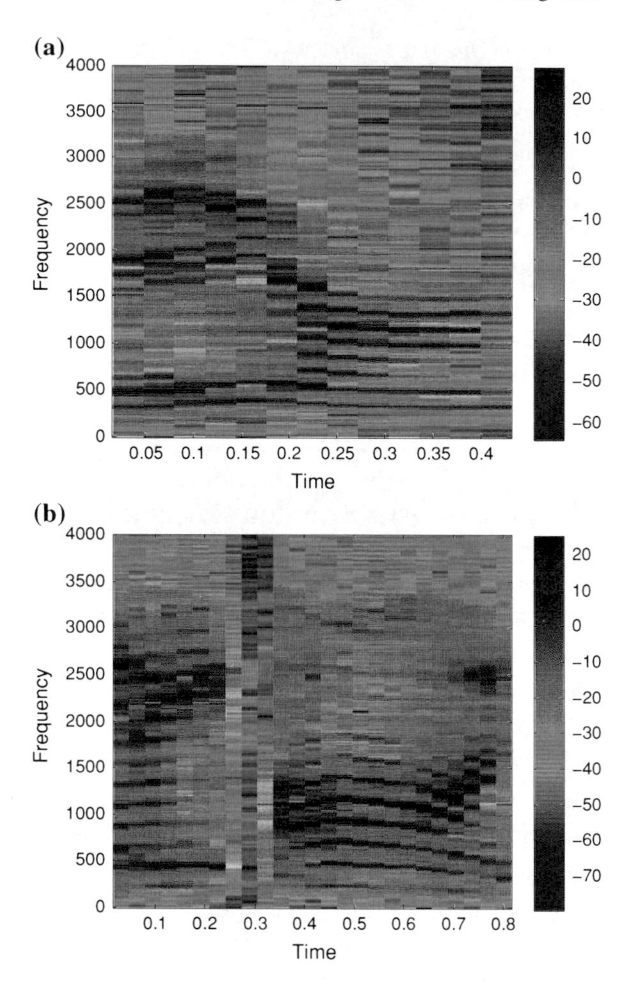

Fig. 4.2 Spectrogram of a user saying "steady on" in a non-angry manner (**a**) and in angry manner (**b**) after being misunderstood by the system. It should be noted that the angry utterance contains more energy in higher frequencies

as depicted here, we can classify and distinguish the user's emotional state on a paralinguistic basis.

Dealing with speech from SDS we usually have to cope with very short utterances that often have command-like style. We suppose, every turn is a short utterance of one prosodic entity. This assumption is to that effect beneficial as dynamic modeling approaches, such as Hidden Markov Models (HMMs) and Dynamic Time Warping (DTW), yield lower performance scores (Schuller 2006), which likewise turned out to be true in our own non-published experiments. In the following, the emotional content of an utterance is thus, due to this short length, summarized as a feature vector of static size. Consequently, we calculate our statistics to account for entire utterances. The process is depicted in Fig. 4.3, which is structured into two consecutive steps. In the first one, an audio descriptor extraction unit processes the raw audio format

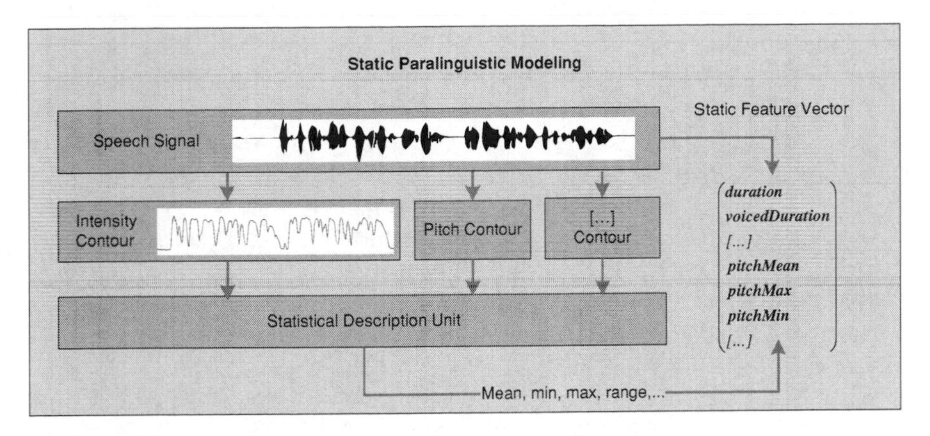

Fig. 4.3 Paralinguistic modeling: the extracted audio descriptor contours are subject to a statistical description unit. The output of the statistical description unit plus basic audio features form the components of the static feature vector representing the entire utterance

and provides speech descriptors. In the second step, various statistics are calculated on both the descriptors and certain subsegments of them.

Paralinguistic Modeling

The paralinguistic information in speech is first modeled by a common feature set that corresponds to state of the art approaches, see e.g., Petrushin (1999); Lee et al. (2001); Liscombe et al. (2005); Pittermann et al. (2009); and Morrison et al. (2007). We will refer to this approach in the following as P_{std}.

An acoustic and spectral analysis extracts relevant features that indicate the emotional state of the speaker. Thereby, the entire utterance is processed using a Gaussian window with 10 ms timeshift.

Pitch

The *pitch* contour from the speech signal is extracted by using autocorrelation as described in Boersma and Weenink (2009) with choosing a pitch range of 75–600 Hz.

Intensity

Taken directly from the speech signal we extract the contour of *intensity*. Taking the square of discrete amplitude values we convert every windowed frame's energy into dB scale relative to the auditory threshold pressure. To avoid any DC offset the mean pressure is subtracted before calculation.

Harmonics-to-Noise Ratio

Moreover, we calculate the *Harmonics-to-Noise Ratio* (HNR) contour, also denoted as "Harmonicity", which is expressed in dB. The HNR estimates the degree of acoustic periodicity in the signal by means of periodicity detection. For example, if 95 % of the energy of the signal is in the periodic part, and 5 % is in the noisy part, the HNR is $10 \cdot \log_{10}(95/5) = 12.8\,\text{dB}$. If the HNR is 0 dB, the energy in the harmonic and in the noise part is equal. We furthermore determine *shimmer*, which is the average difference between two subsequent amplitudes of a signal. It expresses the variation in loudness. To measure the variation in pitch, we further determine the average *jitter* value for the signal.

Formants

Further, we extract five formant center frequencies and search for *formants* up to approximately 3.5 kHz. We apply a pre-emphasis of 6 dB/octave before computing linear predictive coding (LPC) coefficients after Burg as given by Press et al. (1992). We extract formants only for voiced regions as the algorithm yields reasonable results for voiced segments only.

MFCC

Although *MFCCs* are most commonly used in speech recognition tasks they often give excellent performance in anger recognition tasks as well (Nobuo and Yasunari 2007). To obtain MFCC coefficients we perform a filter bank analysis on a Mel-frequency scale and convert the filter values to Mel-frequency cepstral coefficients, which originate from the discrete cosine transform of the filter bank spectrum (in dB). We filter the spectrum into Mel domain units and apply a discrete cosine transformation (DCT), which gives the values of the MFCCs. We place the filter centers in equally spaced intervals of 100 Mel distance and compute a number of 12 MFCC coefficients.

Others

Finally, we determine durational features of the speech signal, such as the *duration* of the utterance and respectively the duration of *voiced* and *unvoiced* parts.

Statistical Description Unit

Based on the calculated descriptor contours of pitch, intensity and harmonicity, as well as the values obtained at a distance of 10 ms, we create a static feature vector representing the entire utterance. The contours describing the pitch, intensity and harmonicity descriptors are modeled by using mean, minimum, maximum, range, and deviation features that are calculated over the entire contour. We further create the first-order derivative for pitch and intensity, also denoted as delta (Δ), to model the change in the contours and in this way the speech signal dynamics.

Feature Set Summary

The employed acoustic and prosodic features that represent a single utterance are as follows:

- **Pitch**, $\boldsymbol{\Delta}$**Pitch**: Mean, minimum, maximum, range, standard deviation.
- **Intensity**, $\boldsymbol{\Delta}$**Intensity**: Mean, minimum, maximum, range, standard deviation.
- **Harmonicity**: Mean, minimum, maximum, range, standard deviation, jitter, shimmer.
- **Formants 1–5**: Mean.
- **MFCC 1–12**: Mean.
- **Others**: Duration of the utterance in seconds, duration of voiced parts, duration of unvoiced parts, ratio of voiced to unvoiced parts.

The feature set comprises 53 acoustic and prosodic features and has been employed in a number of studies, see Herm et al. (2008); Mowafey et al. (2009); and Schmitt et al. (2009a, b).

Extended Paralinguistic Modeling

The previously described modeling, which is similarly used in the literature, covers the most relevant acoustic and prosodic descriptors. However, it can be assumed that this procedure does not mirror all characteristics required for statically modeling emotions in speech. Notably, the dynamics in the signal are only inadequately captured by merely taking into account minimum, maximum, range, and deviation of each respective descriptor. It, thus, neglects details in the dynamics such as higher order derivatives of the contours and further statistical descriptions. In the following, we extend the basic model by introducing further acoustic and prosodic features and by introducing a statistical description unit capturing the dynamics in more detail. We will refer to this approach in the following as P_{ext}.

The resulting audio descriptors can be subdivided into seven groups: *pitch, intensity, formants, MFCC, loudness, spectrals*, and *other*. We calculate pitch in the same manner as in the standard procedure using the pitch algorithm of Boersma and Weenink (2009). To avoid octave jumps in pitch estimation, which frequently leads to wrong pitch values, we additionally post-process a range of possible pitch values using relative thresholds between voiced and unvoiced candidates. Remaining octave confusions between subsegments of a turn are further processed by a rule-based path finding algorithm. In order to normalize for the absolute height of different speakers we convert pitch into the semitone domain using the mean pitch as reference value for a whole turn. As pitch is not defined for unvoiced segments we apply piecewise cubic interpolation and smoothing by local regression using weighted linear least squares.

Intensity and *formants* are calculated in the same manner as previously described in the standard feature set. We further determine the *MFCC coefficients* in analogy to the basic approach, but extend to 16 MFCC coefficients instead of 12 and keep the zero coefficient.

Loudness

Another perceptively motivated measurement is the *loudness* as defined by Fastl and Zwicker (2005). This measurement operates on a Bark filtered version of the spectrum and finally integrates the filter coefficients into a single loudness value in sone units per frame. In contrast to pitch, which is only defined for voiced parts, this measurement is always defined so we obtain a continuous descriptor contour.

Spectrals

Other features drawn from the cepstral representation of the speech signal are the center of spectral mass gravity (spectral centroid) and the 95 % roll-off point of spectral energy. Both features capture aspects related to the spectral slope (also called the spectral tilt) and correspond to perceptual impression of sharpness and brightness of sounds (Fastl and Zwicker 2005). Another measurement drawn from spectral representation is the magnitude of spectral change over time, also known as spectral flux. The more abruptly changes in the spectrum occur, the higher the magnitude of this measurement. Grouping the center of spectral mass, the roll-off point and the spectral flux together these features will be referred to as *spectrals*.

Others

Summarized as *other* features we calculate the HNR as previously described. This measurement is calculated for voiced segments only. Next, we add a single coefficient for the correlation between pitch and intensity as an individual feature. Examining the signal amplitude we calculate the zero-crossing-rate and estimate the average amplitude over the turn. Finally, taken from the relation of pitched and non-pitched speech segments we calculate durational or rhythm-related features such as pause lengths and the average expansion of voiced segments.

Extended Statistical Description Unit

For a more detailed modeling of the acoustic and prosodic descriptors, we introduce an extended statistic unit that provides a detailed description of the respective features. In the first place, the statistic unit derives means, moments of first to fourth order, extrema and ranges from the respective contours. Resulting features are e.g. the standard deviation of the pitch, the average loudness level, the mean of a Mel-frequency cepstral coefficient, the maximum change in spectral flux, the range of variations in bandwidths of a certain formant, the distribution skewness of the intensity level or the minimal level of harmonicity.

Special statistics are then applied to certain descriptors such as pitch and loudness. These descriptors are examined with a *linear regression analysis*. We also include the error coefficient from the analysis as feature in order to have an estimation of linearity of the contours.

Furthermore, pitch, loudness, and intensity are additionally processed by a DCT. By applying DCT to these contours directly, we model their spectral composition. There exist different norms of DCT calculation. We refer to a DCT type III which is defined as:

$$X_k = \frac{1}{2}x_0 + \sum_{n=1}^{N-1} x_n \cos\left[\frac{\pi}{N}n\left(k + \frac{1}{2}\right)\right] \quad k = 0, \ldots, N-1. \tag{4.1}$$

A high correlation of a contour with the lower coefficients indicates a rather slowly moving time behavior while mid-range coefficients would rather correlate with fast moving audio descriptors. Higher order coefficients would correlate with microprosodic movements of the respective curves, which corresponds to a kind of shimmer in the power magnitude or jitter in pitch movement.

A crucial task is the time normalization. In order to exploit the temporal behavior at a certain point of time we append Delta coefficients of *first* (Δ) and *second* ($\Delta\Delta$) order and calculate statistics on them alike. For certain features, we determine the *skewness* and *kurtosis* of the respective value distributions (Joanes and Gill 1998). Skewness represents a measure for the asymmetry of a probability distribution while kurtosis describes its "peakedness". We further determine the *interquartile range* (IQR) as measure for statistical dispersion. It is calculated as difference of the quartiles Q.25 and Q.75 of the respective distributions.

As already mentioned, some features tend to give meaningful values only when applied to specific segments. We, therefore, developed an extended version of the speech-silence detection proposed by Rabiner and Sambur (1975). After having found a range of voiced points we move to the very first and the very last point and search for adjacent areas of relatively high zero-crossing rates. Also any non-voiced segment in between the outer borders is classified into high and low zero-crossing regions corresponding to unvoiced or silent speech segments. Eventually, we calculate features on the basis of *voiced and/or unvoiced sounds* both separately and jointly. In order to capture magnitudes of voiced to unvoiced relations we also compute these quotients as ratio measurements. We apply it to audio descriptors such as intensity and loudness to obtain

- the ratio of mean of unvoiced to mean of voiced points,
- the ratio of median of unvoiced to median of voiced points and
- the ratio of maximum of unvoiced to maximum of voiced points.

Feature Set Summary

In summary, the features representing a single user utterance are the following (new features and feature sets are printed in italic):

- **Pitch**, **ΔPitch**, ***ΔΔPitch***: Mean, median, maximum, minimum, standard deviation, *IQR, linear regression, skewness, kurtosis, DCT coefficients 1–30, ratio unvoiced to voiced* (*mean, median, maximum*).
- **Intensity**, **ΔIntensity**, ***ΔΔIntensity***: Mean, median, maximum, minimum, standard deviation, *IQR, linear regression, skewness, kurtosis, DCT coefficients 1–30.*
- ***Intensity voiced points/unvoiced points***: Mean, median, maximum, minimum, standard deviation, *IQR, linear regression, skewness, kurtosis.*
- *Correlation of Pitch and Intensity.*
- **Formants 1–5**, **ΔFormants 1–5**, ***ΔΔFormants 1–5***: Mean, minimum, maximum, range, standard deviation; *bandwidth mean, bandwidth median, bandwidth maximum, bandwidth minimum, bandwidth standard deviation, bandwidth range.*
- **HNR**: Mean, standard deviation, maximum.
- ***Loudness***, ***ΔLoudness***, ***ΔΔLoudness***: Mean, median, maximum, minimum, standard deviation, IQR, linear regression, skewness, kurtosis, DCT coefficients 1–30, ratio unvoiced to voiced (mean, median, maximum).
- ***Loudness voiced points/unvoiced parts***: Mean, median, maximum, minimum, standard deviation, *IQR, linear regression, skewness, kurtosis.*
- **MFCC 0–16**, **ΔMFCC 0–16**, ***ΔΔMFCC 0–16*** (respectively of entire utterance, only voiced segments and only unvoiced segments): Mean, maximum, minimum, standard deviation.
- ***Spectrals***: ***spectral flux***, ***role-off point respectively for entire utterance***, ***voiced and unvoiced parts***: Mean, maximum, minimum, standard deviation, range.
- **Time features**: Duration of the utterance in seconds, duration of voiced parts, duration of unvoiced parts, ratio of voiced to unvoiced parts, ratio of pause to entire utterance.

We obtain 1,477 features in total. For related studies employing the presented paralinguistic modeling, cf. Schmitt et al. (2010b, c, d) and Polzehl et al. (2009, 2010a, b).

4.1.2 Linguistic Emotion Recognition

Linguistic emotion recognition models the information given by the manual transcription of the spoken words or the automatic transcriptions obtained from the ASR of the user's utterances. We will refer to linguistic models as **L**. For automatic emotion recognition, the employment of manual transcriptions does not yield realistic conditions, since they are not available at runtime. Such transcriptions are only relevant to provide an upper baseline.

Some linguistic modeling approaches use static, hand-crafted word lists, see e.g., Pittermann et al. (2009) and Pittermann and Schmitt (2008). The assumption hereby is that certain words co-occur with a specific emotion class, e.g., the word "like" with the emotion class "happy". The disadvantage of this procedure is that this definition is somewhat arbitrary and a large proportion of words in the list does not necessarily correlate with a specific emotion class label. In this section, we explore statistical techniques modeling a linguistic dependency of an utterance's textual content with the class label assigned by the expert annotator. The advantages of this approach are twofold: first, statistical modeling allows for a rapid adaptation of the linguistic model to the respective application domain by simply re-training the model. Second, instead of relying on a presumed and probably false dependency between a word and an emotion class, the true dependency of the textual content with the actual emotion class is exploited.

We investigate the performance of word modeling for emotion recognition using four different feature spaces, i.e., Bag-of-Words (BOW), Term Frequency (TF), Term Frequency-Inverse Document Frequency (TF-IDF), and the Self-Referential Information (SRI). BOW, TF, and TF-IDF are frequently used in information retrieval tasks, e.g., document classification (Huang et al. 2001).

BOW

BOW builds up a feature space by collecting all words from a dataset and by registering the words contained in an individual document. Our BOW model contains 1 or 0 for present or absent words, respectively. As a result, word vector representations can be very high-dimensional and consequently the feature spaces are sparsely populated.

TF; TF-IDF

TF refers to a similar method. Instead of marking absence and presence, normalized word counts are registered in the vector space. TF-IDF weights TF by the inverse of the document frequency, i.e., the frequency of documents containing an individual word, see also Sect. 2.2.4.

SRI

Departing from the concept of relative entropy between two probability mass functions, we calculate the information of a word with respect to an emotion class. Let $w \in W$ be a word out of a vocabulary, $\varepsilon \in E$ an emotion out of all target emotion classes and $P(\varepsilon)$ the prior probability of an emotion. The SRI about an emotion class, given a posterior probability that a certain word implies a certain emotion, can be estimated by:

$$\mathrm{SRI}(\varepsilon, w) = \log \frac{P(\varepsilon|w)}{P(\varepsilon)} \tag{4.2}$$

It should be noted that a negative SRI value for a word implies that the word makes an emotion class less likely, whereas a positive value implies that the emotion class is more likely. To estimate the "emotional content" of an *utterance* we sum up the word- and class-specific SRI values and decide for the class of maximum SRI sums.

Emotional Salience

Departing from SRI we calculate the *mutual information* (MI) (Manning et al. 2008) of a word and the emotion classes. MI measures the dependency between two random variables, here the words and the emotion classes. Statistical independence of the two probabilities would result in an MI of zero. Consequently the higher the MI value, the stronger the correlation with the class labels. In this context, MI is also denoted as *Emotional Salience* (Lee et al. 2005). Let k be the number of classes, then MI, i.e., Emotional Salience, is defined as:

$$\mathrm{sal}(w) = \mathrm{MI}(E, W = w) = \sum_{j=1}^{k} P(\varepsilon_j|w) \cdot \log \frac{P(\varepsilon_j|w)}{P(\varepsilon_j)} \tag{4.3}$$

A word obtaining a high Emotional Salience score implies high emotional content. Thereby, the Emotional Salience does not express to which emotion class a correlation exists. For modeling Emotional Salience in a dataset, we calculate the activation feature proposed by Lee et al. (2005), which weights the SRI word summation with the prior class probabilities at the exchange level.

Table 4.1 presents ten examples from the most salient words for the three databases SC-BAC, MOB, and AIBO, which are introduced later in this chapter.

4.2 Dialog-Based Emotion Recognition

The previously described paralinguistic and linguistic modeling approaches are restricted to the information available from a speech sample, i.e., a recorded single user utterance. However, in a spoken dialog between an SDSand a user further

Table 4.1 Salience of words obtained from SC-BAC, MOB and AIBO, cf. Sect. 4.3.1

Class	SC-BAC Word	Salience	MOB Word	Word Salience	AIBO	Salience
ANGER	Wrong	1.3	dämlicher (*stupid*)	1.2	Schluss (*finish*)	1.5
ANGER	Operator	0.8	Teuer (*expensive*)	0.6	Stoppen (*to stop*)	1.0
ANGER	Person	0.7	Doch ⟨⟨*exasperation*⟩⟩	0.5	Aufhören (*to end*)	0.7
ANGER	Please	0.7	Warum (*why*)	0.5	Faul (*lazy*)	0.5
ANGER	Support	0.5	Falsch (*wrong*)	0.2	Endlich (*finally*)	0.3
NONANGER	Correct	0.7	Korrekt (*right*)	0.7	Brav (*good/obedient*)	0.5
NONANGER	Okay	0.3	Einfach (*simple*)	0.6	Fein (*fine*)	0.5
NONANGER	Right	0.2	Danke (*thanks*)	0.3	Schön (*nice*)	0.4
NONANGER	Ready	0.1	Okay	0.2	gut (*good*)	0.4
NONANGER	Connected	0.1	Bonuspunkte (*bonus points*)	0.1	Okay	0.3

information may be derived that may contribute to the reliable estimation of the user's emotional state. Thus, we will first examine and model the dependency of the interaction and resulting emotions. Second, we will analyze the occurrence of emotions in the dialog context and model the history of the user's emotional state. Third, SDS-related problems for deploying emotion recognition are discussed and resolved.

4.2.1 Interaction and Context-Related Emotion Recognition

The emotion annotation process for the Broadband Agent brought to light that some users react angry or at least displeased right away once realizing that they are served by a machine. Others are dissatisfied and show anger already at the beginning due to a problem with a product or breakdown of a service. Apart from these cases, we assume that emotions in interactions with SDS do not emerge causelessly but depend to a certain degree on external factors. For example, according to our observations, the poor system behavior and to some degree poor system design are the most important triggers for anger. Knowledge about interaction patterns between the system and the user may, thus, contribute decisively to a reliable recognition of emotions. However, cause and effect in this field have only been tentatively analyzed, with some few studies approaching this subject. López-Cózar et al. (2008) observed increased anger when the system repeatedly posed the same question after a malfunction of the ASR. The study reports a performance increase by 1.3–3.0 % when employing dialog act information in a *negative* versus *non-negative* classification task. The joint classifier using lexical, prosodic, acoustic, and dialog act information reaches a maximum performance of 92.23 %. Similar results have been reported by Ai et al (2006). Incorporating information about system–user performance increases the classification accuracy by relatively 2.7 %. Liscombe el al. (2005) reported a 1.2 % performance increase when employing dialog act features. All three studies use merely selective information from the system–user interaction. Here, dialog acts are often the only employed feature, but they bring along the necessity of hand annotation. A comprehensive set of interaction parameters as described in Sect. 3.2.1 that can be derived automatically from system logs has not been evaluated on the task of emotion recognition. Further, a feature analysis, i.e., a feature ranking presenting the most important parameters, is vital for an understanding of the interplay between system–user interaction and emotions.

Correlations Between the Negative Emotional State and Interaction Parameters

For assessing the correlation between emotions and interaction parameters, we perform a correlation analysis using Spearman's ρ as described by Bartholomew (2002). The annotated emotional state has been mapped to an ordinal numerical scale, whereas the mapping is as follows: NONANGER = 1, SLIGHTANGER = 2 and STRONGANGER = 3, i.e., the target variable ε represents the degree of anger. Correlations have been determined on the 21,432 utterances from SC-BAC and the 4,243 utterances from LG-FIELD that contained an annotation $\in$ {NONANGER, SLIGHTANGER,STRONGANGER}, i.e., garbage turns have been excluded. Input variables for this analysis are the *numerical* interaction parameters. The top 20 correlations of interaction parameters with the target variable ε are depicted in Table 4.2.

We can see already weak dependencies when statistical independence is assumed. It is interesting to note that the correlations in SC-BAC are generally lower compared to LG-FIELD. This might be due to the distribution of NONANGER versus ANGER, which is very skew toward NONANGER classes in SC-BAC. In contrast to this, LG-FIELD contains a higher degree of ANGER. Analyzing the correlation scores of the isolated parameters we can observe a number of dependencies. The duration of the utterance seems to play an important role, since the parameter UTD positively correlates with the degree of anger in both corpora. That means that longer utterances frequently imply angry users. Apparently, user anger increases in LG-FIELD with progressing dialog. It seems that users in general get more annoyed the longer the interaction lasts, which can be seen at the positive correlation of the cumulative interaction parameters #REPROMPTS, #USERTURNS, #ASRSUCCESS, #SYSTEMTURNS with ε. In contrast to this, users in SC-BAC are more strongly influenced by malfunctions in the immediate context of the interaction, since higher correlations can be reported for contextual parameters. For both corpora it can be said that ASR performance influences the negative emotional state. While malfunctions provoke anger, error-free ASR operation leads to less angry users, which can be seen at the negative correlations of {#}ASRSUCCESS, %ASRSUCCESS, ASRCONFIDENCE and {MEAN}ASRCONFIDENCE in both corpora. For example, the better the ASR recognizes the user in a dialog (%ASRSUCCESS), the less anger may be observed.

Table 4.2 Top 20 rank correlation coefficients (Spearman's ρ) of numerical interaction parameters with the negative emotional state in SC-BAC and LG-FIELD

SC-BAC		LG-FIELD	
{#}TIMEOUTS_ASRREJ	0.143	#REPROMPTS	0.348
{#}REPROMPTS	0.138	#USERTURNS	0.33
REPROMPT?	0.136	TURNNUMBER	0.329
UTD	0.134	%REPROMPTS	0.323
{#}UNEXMO	0.131	#ASRSUCCESS	0.315
ROLEINDEX	0.113	#SYSTEMTURNS	0.304
%ASRSUCCESS	−0.108	#BARGE- INS	0.300
%TIMEOUTS_ASRREJ	0.108	#SYSTEMQUESTIONS	0.295
%REPROMPTS	0.103	{#}REPROMPTS	0.274
{MEAN}ASRCONFIDENCE	−0.101	#ASRREJECTIONS	0.266
%UNEXMO	0.100	#TIMEOUTS_ASRREJ	0.262
{#}ASRSUCCESS	−0.099	UTD	0.247
WER	0.099	ROLEINDEX	0.205
#OPERATORREQUEST	0.099	#TIME- OUTPROMPTS	0.202
%OPERATORREQUEST	0.099	REPROMPT?	0.194
#REPROMPTS	0.097	%ASRREJECTIONS	0.191
OPERATORREQUEST?	0.097	%ASRSUCCESS	−0.179
%TIME- OUTPROMPTS	0.097	{#}ASRSUCCESS	−0.177
#TIMEOUTS_ASRREJ	0.096	ASRCONFIDENCE	−0.173
#UNEXMO	0.092	%TIMEOUTS_ASRREJ	0.162

All correlations are highly statistical significant at the 0.001-level

Although it can be assumed that angry users tend to interrupt the system prompt, i.e., they barge in, no empirical evidence can be found in both corpora, due to the absence of *relative* barge-in parameters in the top-ranked correlations.

Feature Set Summary

For modeling the emotional state of the user at turn t, we employ the following interaction parameters as features in a 53-dimensional static feature vector (for definitions of the parameters cf. Sect. 3.2.1):

ASR ASRCONFIDENCE, ASRRECOGNITIONSTATUS, BARGE- IN?, EXMO, GRAMMAR, MEANASRCONFIDENCE, MODALITY, TRIGGERED GRAMMAR, UTD, UNEXMO?, WPUT, #BARGE- INS, #ASRREJECTIONS, #ASRSUCCESS, #TIMEOUTPROMPTS, #TIMEOUTS_ASRREJ, #UNEXMO, %BARGE- INS, %TIMEOUTS_ASRREJ, %ASRSUCCESS, %TIMEOUTPROMPTS, %UNEXMO, {MEAN} ASRCONFIDENCE, {#}BARGE- INS, {#}ASRREJECTIONS, {#}ASRSUCCESS, {#}TIMEOUTPROMPTS, {#}TIMEOUTS_ASRREJ, {#}UNEXMO;

SLU HELPREQUEST?, OPERATORREQUEST?, SEMANTICPARSE, #OPERATOR REQUESTS, #HELPREQUESTS, %OPERATORREQUESTS, {#}HELPREQUESTS, {#}OPERATOR REQUESTS;

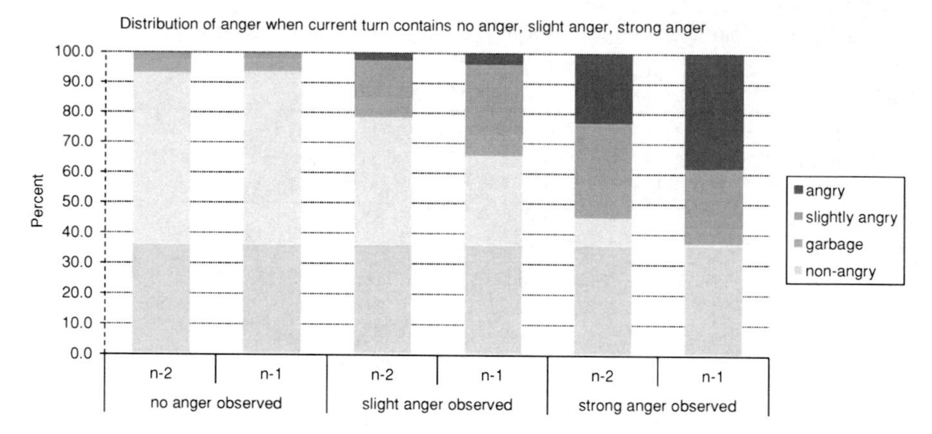

Fig. 4.4 Emotional history of all exchanges from a field corpus. Considered are the two previous dialog turns of the current turn. For example, when we observe anger in the current turn (see last bar), the likelihood that the user has already been angry in the previous turn $n-1$ is 38 %. For details cf. Table B.1.1 in the appendix. The statistic is derived from all utterances of the SC-BAC corpus (cf. Sect. 4.3.1)

DM ActivityTrigram, ActivityType, Activity, DD, Loop name, Prompt, Reprompt?, RoleIndex, RoleName, WPST, #Exchanges, #Reprompt, #SystemQuestions, #SystemTurns, #UserTurns, %Reprompt, {#}Reprompt, {#}System Questions.

Explicit lexical information as contained in the Utterance parameter has been omitted since it is already implicitly included in the linguistic model. The interaction- and context-related feature vector consists of n dimensions and may be applied on any discriminative classifier such as kNN, SVM, ANN, and decision trees.

We will refer to the interaction- and context-related model as **I and C**.

4.2.2 *Emotional History*

An important factor has been neglected so far when dealing with emotion recognition in SDS: users do not show an emotion out of the clear blue sky. A certain "history" can be observed, which holds true particularly for the emergence of anger as depicted in Fig. 4.4.

It can be seen in the first two bars of the chart that it would be highly unlikely that a user who is non-angry in the current dialog turn had been slightly angry, or intensely angry for that matter, in the two previous turns. By contrast, it is interesting to analyze the anger history of turns where the user showed slight or perhaps even intense anger. For example, if we observe that the user is slightly angry in the 5th dialog turn, the likelihood that the user has already been slightly angry in the two previous turns, i.e., the 4th and the 3rd turn are 24 and 13 %, respectively. In other words, when a user is angry in his current turn we have a very high probability that he will be angry in the next turn as well. The role of garbage turns is also striking. In this context, garbage turns are turns where the user did engage in cross-talk or when other background noise and non-speech events such as coughing or sneezing have been recorded. When comparing non-angry and angry turns it is interesting to see that angry turns were more frequently preceded by garbage turns. The likelihood of having had a garbage turn two steps prior to a hot anger amounts to 11 % (second last bar). Such a garbage turn inevitably leads to an ASR error which then causes the system to re-prompt the question.

If we analyze the emotional state of the user prior to the currently observed angry turn in the analyzed speech corpus, we observe sequences that might look as follows when considering the three earlier turns:

ANA, NAN, NAA etc.

It should be noted that "A" stands for an angry user turn and "N" stands for a non-angry user turn. Obviously, the current emotional state has a certain history of previous anger and frustration. On the other hand, when looking at the three prior turns of a non-angry utterance, we observe much less angry turns:

NNN, NAN, NNN etc.

The likelihood of observing a specific emotion can be statistically implemented by using HMMs. The basic idea of HMMs has already been described in Sect. 2.1.1.

Let $\varepsilon(t) \in E$ be the user's emotional state at dialog turn t, where E are all k possible emotion classes. Further, let $\Omega = \{\varepsilon(t_1), \ldots, \varepsilon(t_{n-2}), \varepsilon(t_{n-1})\}$ be an observation sequence of emotional states ε_i occurring in the respective user turns t_i prior to the currently observed user turn t_n. Then, we can discriminate between k different types of sequences Ω_E, which are observations leading to an emotion ε. The probability of the user's emotion at turn t_n belonging to the class ε can be modeled by an ensemble of k HMMs, where

$$M_\varepsilon = (\mathcal{S}, \mathcal{A}, \pi, \Omega_\varepsilon, \mathcal{B}) \tag{4.4}$$

with the respective state transition probabilities

$$\mathcal{A} = \begin{pmatrix} a_{11} & a_{12} \\ a_{21} & a_{22} \end{pmatrix} \tag{4.5}$$

and the observation symbol probability distribution

$$\mathcal{B} = \begin{pmatrix} b_{11} & b_{12} \\ b_{21} & b_{22} \end{pmatrix} \tag{4.6}$$

as well as the initial state distribution

$$\pi = (P_1 \ P_2)\,. \tag{4.7}$$

The values of the matrices $\mathcal{A}$, $\mathcal{B}$ and π are respectively obtained by training M_ε with sequences of Ω_ε. $\mathcal{S}$ denotes the number of hidden states in the model, which is assumed to be 2 in this setting.

To determine the most likely emotion class $\hat{\varepsilon}$ given an observation sequence Ω at the current turn t_i we determine the production probabilities P of all k

$$\hat{\varepsilon} = \arg\max_{M_E} P_{M_E}(\Omega | E), \tag{4.8}$$

i.e., we pick the emotion class where the corresponding HMM yielded the highest production probability. We will refer to the history-related model as **H**.

4.2.3 Emotion Recognition in Deployment Scenarios

In previous work, it is assumed that the recognition of all emotion classes is of equal importance. Consequently, the costs of misclassification are considered equal for all classes. This assumption does not hold true for the deployment of an emotion recognition system in a real-life SDS, which is best explained by a binary anger detection example.

A discriminative binary classification task such as anger recognition aims to distinguish between two classes. By definition, one class is denoted as "positive", the other as "negative". In the assessment of a binary classifier, it is of relevance how many true and false positives as well as how many true and false negatives are generated, see Sect. 2.1.2.

Porting these numbers to the anger recognition domain, the metrics TP, TN, FP, and FN signify, how many utterances where the user has

- *been angry* have been *correctly* classified by the classifier? (TP)
- *been angry* have been *mistakenly* classified as *non-angry*? (FN)
- *not been angry* have been *correctly* classified as *non-angry*? (TN)
- *not been angry* have *mistakenly* been classified as *angry*? (FP)

The use of emotion detection in SDS implies that a certain action follows once the emotional state is determined. This may be an adaptation of the strategy or an escalation to a live agent, if e.g., anger is spotted. Let us assume a classifier spotting negative emotions would be deployed in a field SDS and the action taken would be an escalation of the user to a live agent, once a sequence of negative emotions are spotted. Then it appears obvious that it is less serious if the classifier yields a high number of FN than a high number of FP: if a user turn is classified as non-angry although the user is angry (FN), the dialog system would not behave differently than when no anger detection is deployed. A completely different scenario would occur when a non-angry user is mistakenly classified as angry (FP): the SDS would transfer the user to an operator although there is no need of doing so. In order to prevent a high number of FPs we apply a cost-based approach.

To make emotion classification cost sensitive, we use the MetaCost algorithm (Domingos 1999) to our classifier. MetaCost uses the confidence scores of the underlying classifier in order to choose the class with the minimal risk. In this context, the conditional risk $R(i|x)$ is defined as

$$R(i|x) = \sum_j P(j|x)C(i, j). \tag{4.9}$$

The risk is the expected cost of classifying sample x as belonging to class i. The probability P is the confidence of the underlying classifier that sample x belongs to class j. The cost matrix $C(i, j)$ defines the costs for correct and wrong predictions. It contains penalty values for each classifier decision when predicting the class of an utterance. Normally, each correct decision is weighted with 0 (no penalty, since it was a correct decision) and each misclassification is weighted as 1. Altering these weights causes the classifier to favor one class at the charge of the other class.

It should be noted that the costs described here depict virtual costs. The true costs of misclassifying an emotion can hardly be quantified or may depend on a variety of factors, such as costs for live agents, availability of live agents, per-minute-costs for the IVR system, or the "degree of user annoyance" that is caused by wrong classification, etc.

4.3 Evaluation

Nearly all studies on emotion recognition, and notably the ones on anger recognition, are based on single corpora. This makes a generalization of the results difficult. Our aim in this book is to compare the performance of different modeling techniques and features when trained and tested on different corpora. All selected databases account for real-life conditions, i.e., they have background noise, recordings include cross- and off-talk, speakers are free in the choice of words and do not enunciate as clearly as trained speakers do.

Common to all IVR databases is the *virtual absence of emotions other than anger*, i.e., happiness, boredom etc. do extremely rarely occur, making their detection irrelevant for the further dialog progress. The focus in this evaluation thus lies in the detection of angry user utterances. In cases where the databases have been annotated with different degrees of anger, all anger classes are summarized as class ANGER. The neutral emotional state is summarized with the remaining emotions as class NONANGER.

In order to compare the performance of our models for the different databases, we calculate classification success using six evaluation scores: f1-measure, class-wise F-measures, as well as class-wise precision and recall measures. In contrast to most related studies, we omit the use of the *accuracy* measurement with good cause: given a skew class distribution, the accuracy measure overestimates if the model of the majority class yields better results than the models for the non-majority classes. Consequently, an accuracy measure may misleadingly indicate a good performance of a classifier, although poor performance scores are obtained for minority classes. Comparisons between results of studies that use accuracy measures are thus often biased and, in some cases, may even be invalid. We thus primarily employ the f1-measurement. It is defined as the (unweighted) average of all class-specific F-measures, which expresses the classifier's performance on predicting a specific class. The F-measure itself accounts for the harmonic mean of both precision and recall of a given class, cf. Sect. 2.1.2.

4.3.1 Corpora

Four real-life corpora from a large number of speakers are employed to assess the performance of the models, three IVR corpora and one Wizard-of-Oz corpus.

The *SC-BAC* database originates from a US-American portal designed to solve Internet-related problems jointly with the caller. It helps customers to recover Internet connections, reset lost passwords, cancel appointments with service employees or reset lost e-mail passwords. If the system is unable to help the customer, the call is escalated to a human operator. The annotation scheme and process has been described in Sect. 3.2.2. Three labelers divided the corpus into *angry*, *annoyed* and *non-angry* utterances. The final label was defined based on majority voting resulting in 90.2 % neutral, 5.1 % garbage, 3.4 % slightly angry, and 0.7 % very angry utterances. 0.6 % of the samples in the corpus were eliminated because all three raters had different opinions. While the number of angry and annoyed utterances seems very low, 429 calls (i.e. 22.4 % of all dialogs) contained annoyed or angry utterances. A more detailed description of the corpus can be found in Schmitt et al. (2010g). The annotation process has been described in Sect. 3.2.2.

The *MOB* database contains about 21 h of recordings from a German voice portal. Customers call in to report on problems, e.g., with the phone connection. The callers are being preselected by an automated voice dialog before they are passed to an agent. The data can be subdivided into 4,683 dialogs, averaging 5.8 turns per dialog. For each turn, three labelers assigned one of the following labels: *not angry*, *not sure*, *slightly angry*, *clear anger*, *clear rage* or marked the turns as *non applicable* when encountering garbage. A more detailed description of the corpus can be found in Burkhardt et al. (2009).

The *LG-FIELD* corpus originates from recordings of the CMU Let's Go Bus information system deployed by Carnegie Mellon University, Pittsburgh, to automatically serve customers of the Port Authority over telephone. Let's Go provides tailored bus schedules on request. The corpus has been annotated by a single rater, cf. 3.2.2, on a 5 + 1 class rating scheme comparable to the scheme of the MOB database.

The annotated labels have been *friendly*, *neutral*, *slightly angry*, *angry*, *very angry* and *garbage*.

The German *AIBO* database consists of children interacting with the AIBO robot dog. A number of 51 children (age 10–13) were recorded in a Wizard-of-Oz scenario. The children were given the task to navigate the robot through a certain course of actions using voice commands. When the robot reacted disobediently, it provoked emotional reactions from the children. The data amounts to 9.2 h of 16 bit/16 kHz speech recordings in total. Five labelers annotated the utterances with respect to 10 emotion-related target classes. A more detailed description of the corpus can be found in Steidl (2009).

In order to be able to compare results of all corpora, we matched the conditions of all databases to a binary data set consisting of ANGER and NONANGER utterances. In SC-BAC we collapsed *slightly angry* and *very angry* to ANGER and created test and training sets according to the 40/60 split. The resulting set consists of 1,560 NONANGER and 1,012 ANGER turns. Following the extension of Cohen's kappa for multiple labelers by Davies and Fleiss (1982), we obtain a value of $\kappa = 0.63$ on this subset, which corresponds to substantial inter-labeler agreement (Landis and Koch 1977). The average turn length after eliminating initial and final pauses is approximately 0.8 s.

The labels of the MOB database were mapped onto two cover classes by clustering according to a threshold over the average of all voters' labels as described by Burkhardt et al. (2009). Finally, our experimental set contains 1,951 ANGER turns and 2,804 NONANGER turns, which corresponds approximately to a 40/60 split of anger/non-anger distribution. The inter-labeler agreement results in $\kappa = 0.52$, which represents moderate agreement. The average turn length after removing initial and final pauses is 1.8 s.

In LG-FIELD, the classes *friendly* and *neutral* have been collapsed to NONANGER, the classes *slightly angry*, *angry* and *very angry* have been merged to ANGER. The subset employed for the experiments amounts to 1,090 NONANGER and 934 ANGER turns. Since only one rater annotated the corpus, no corpus-wide κ value can be given. The average turn length amounts to 1.65 s, when removing initial and final pauses.

The AIBO database has been mapped to a binary division between negative (NEG), subsuming $touchy$, $angry$, $reprimanding$, and $emphatic$ labels, and non-negative (IDL) utterances, subsuming all other classes, as described in (Steidl et al. 2005). Recordings were split into chunks by syntactic-prosodic criteria. For the present experiments we chose a subset of 26 children,[1] which results in 3,358 NEG and 6,601 IDL chunks corresponding to a 33/66 split. The inter-labeler agreement results in $\kappa = 0.56$.

Details of all four corpora are listed in Table 4.3. While the three IVR databases contain different degrees of anger expression in the Anger class, the AIBO database also subsumes other emotion-related states. Thus, more diverse patterns in the AIBO

[1] This set corresponds to the AIBO *chunk train set* used in the INTERSPEECH 2009 emotion challenge (Schuller et al. 2009).

Table 4.3 Database comparison of SC-BAC, MOB, LG-FIELD and AIBO

Database	SC-BAC	MOB	LG-FIELD	AIBO
Description	SC Broadband	Mobile	Let's Go Field	Aibo Robot
Language	English	German	English	German
Type	IVR	IVR	IVR	Wizard-of-Oz
Domain	Automated technical troubleshooting	CRM on mobile topics	Bus schedule information	Children directing robot
Number of dialogs in total	1,911	4,682	328	–
Duration in total	10 h	21 h	2.2 h	9.2 h
Average number of turns/dialog	11.88	5.7	26	–
Number of raters	3	3	1	5
Speech quality	Narrow-band	Narrow-band	Narrow-band	Wide-band
Subsets for evaluation				
Number of speakers	417	683	134	26
Number of turns	2,328	4,515	2,024	9,959
Number of words in total	3,709	11,812	17,275	26,157
Vocabulary size	286	1,179	n.a.	901
Perplexity	40	233	n.a.	78
Average number of words per turn	1.6 ± 3.5	2.6 ± 3.7	3.3 ± 2.3	2.7 ± 1.7
Number of ANGER turns	1,012	1,951	934	3,358

(continued)

Table 4.3 (continued)

Database	SC-BAC	MOB	LG-FIELD	AIBO
Description	SC Broadband	Mobile	Let's Go Field	Aibo Robot
Number of NONANGER turns	1,560	2,804	1,090	6,601
Average utterance length in seconds	0.84	1.8	1.65	0.87
Average duration of ANGER in seconds	1.87 ± 0.61	3.27 ± 2.27	2.14 ± 1.71	0.87 ± 0.51
Average duration of NONANGER in seconds	1.57 ± 0.66	2.91 ± 2.16	1.54 ± 1.32	0.87 ± 0.62
Cohen's extended kappa	0.63	0.52	–	0.56
Evaluation Level				
Human	X	–	–	–
Paralinguistic (P_{std})	X	X	X	–
Extended paralinguistic (P_{ext})	X	X	–	X
Linguistic (L)	X	X	–	X
Interaction and context (I and C)	X	–	X	–
History (H)	X	X	–	–
Fusion I: $P_{ext} + L$	X	X	–	X
Fusion II: $P_{std} + I$ and C	X	–	X	–
Fusion III: $P_{std} + H$	X	X	–	–

ANGER class can be expected. In order to facilitate formal comparisons, we will refer to the NEG and IDL classes in the AIBO database as ANGER and NONANGER classes and consider the given chunks as corresponding to turns. Further, all samples from the selected databases were presented to the labelers chronologically and independently. This way, the history of a turn being part of a dialog course was known to the labelers, i.e., the decision about the label is also based on the dialog context. The labelers of the IVR databases SC-BAC, MOB, and LG-FIELD were familiar with the respective voice portals and linguistic emotion theory. The labelers of the AIBO database were advanced students of linguistics. Rating the turns or chunks, acoustic and linguistic information processing happened simultaneously, i.e., all stimuli were given in audible, not written form.

4.3.2 Human Performance

Pushing the error rate to zero and obtaining a recognition rate of 100 % for all classes is usually the ultimate goal in classification tasks. However, the attainability of this error-free classification mainly depends on the characteristics of the task, the data, the classification algorithms and the features that are employed. Furthermore, with an increasing number of classes involved in the task, the risk of confusing one class with another rises and the performance tends to degrade. In cases where patterns overlap and where they are neither linearly, nor nonlinearly separable, the aim of reaching error-free classification is out of scope. An overlap of feature spaces usually adheres to most speech-based classification tasks.

Informally the performance also depends on the "ease" of the task. A popular procedure to assess this "ease" is the establishment of a human baseline. The central question that is posed is: "Can automatic classification come close or even exceed the performance of a human rater?"

Discriminating between the genders of speakers is straight-forward for both humans and the statistical classifier. For automatically discriminating between genders, even a single rule can be applied to linearly separate male and female which already delivers satisfying performance scores: Abdulla and Kasabov (2001) use a pitch threshold of 160 Hz. The distinction of speaker age turns out to be more complex, for that reason alone that age has to be determined on a continuous scale and due to the fact that age characteristics are less obvious. Even the task whether a speaker is senior ($\geq$60) or non-senior ($<$60) reaches low agreement between human raters. It could be further shown that with decreasing length of a speech sample, the distinction of age is increasingly difficult, see e.g., Schmitt et al. (2010d). While it is much easier for a machine learning algorithm to separate patterns that are clearly separable by obvious features, such as in speech-based gender recognition, it is to date virtually impossible to reach 100 % accuracies in discriminating between intoxicated and non-intoxicated speakers, see e.g., Ultes et al. (2011b).

Discriminating between three emotion classes in noisy, low-quality telephone data can be considered as challenging task. In the following, the human performance on

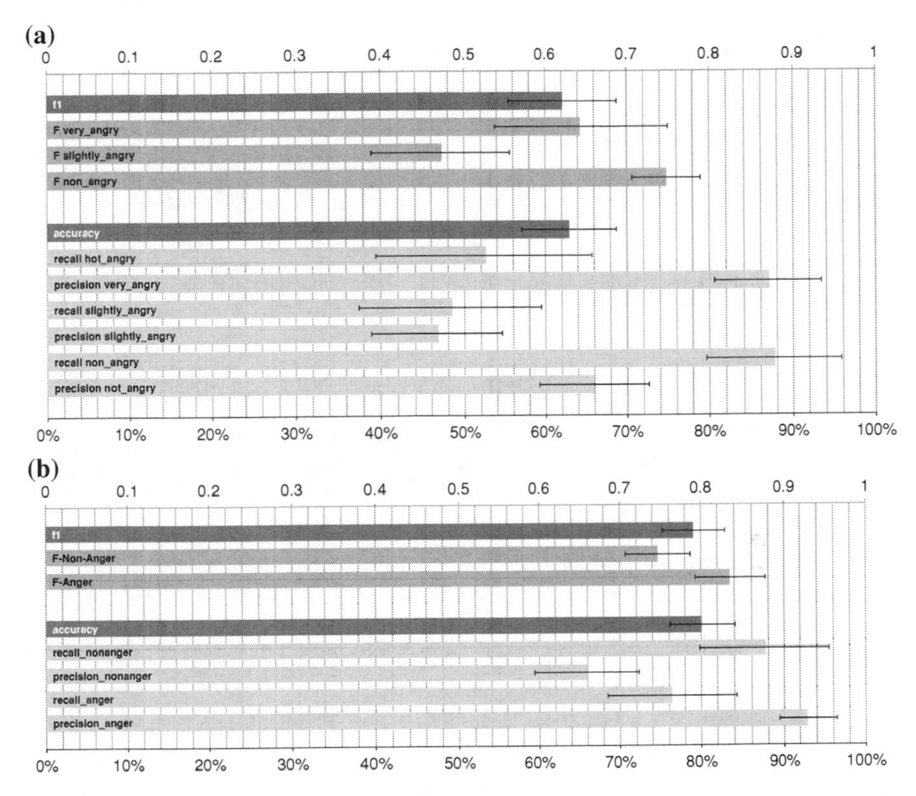

Fig. 4.5 Human performance on SC-BAC: 17 raters respectively annotated 33 non-angry, 33 slightly angry, and 33 very angry utterances from the corpus. Results are depicted in (**a**). For comparison reasons with later experiments, the classes slightly angry and very angry are collapsed to angry in (**b**)

detecting anger in the SC-BAC dataset is assessed. 17 students of computer science have been asked to listen to 99 samples from the corpus consisting of 33 NONANGER, 33 SLIGHTANGER, and 33 STRONGANGER utterances. All participants have been of foreign non-native origin, however, with excellent English language skills. The set employed in the study has been randomly selected. The participants were provided with five examples per class that could be used throughout the rating for comparison. Thereby, we aimed to establish comparable conditions of a trained classifier and the raters.

The results of the study are depicted in Fig. 4.5 and in Table 4.4. For comparison reasons to the binary classification task, the classes SLIGHTANGER and STRONGANGER have likewise been mapped to a more general ANGER class. Depicted are mean values averaged over all raters.

In average, the test persons categorized the samples with $f1 = 0.62$. When looking at the F-measures of the single classes it becomes visible that the human performance among all three classes strongly differs. The best score could be

Table 4.4 Human performance on SC-BAC

Database	Class	Recall (%)	Precision (%)	F-measure	f1-measure
SC-BAC (3 classes)	NONANGER	87.7	65.9	0.79	0.62
	SLIGHTANGER	48.5	46.8	0.47	
	STRONGANGER	52.6	87.0	0.64	
SC-BAC (2 classes)	NONANGER	87.7	65.9	0.75	0.79
	ANGER	76.4	92.9	0.84	

achieved in recognizing NONANGER samples with $F_{\text{NONANGER}} = 0.75$. From all NONANGER utterances nearly all could be identified (cf. $\text{recall}_{\text{NONANGER}}$) while only two-thirds of the selected samples actually belonged to the group NONANGER (cf. $\text{precision}_{\text{NONANGER}}$). The remaining one-third are SLIGHTANGER and STRONGANGER samples that have been mistakenly assigned to the NONANGER category. Obviously, the SLIGHTANGER category has been the most difficult to identify with $F_{\text{SLIGHTANGER}} = 0.47$). This may be due to the fact that it ranges in between the two extremes STRONGANGER and NONANGER as the acoustic characteristics seem to overlap with the neighboring classes, making a decision difficult. In average, one half of the SLIGHTANGER samples have been identified correctly (cf. $\text{recall}_{\text{SLIGHTANGER}}$). Only one half of the selected samples have been correct (cf. $\text{precision}_{\text{SLIGHTANGER}}$), whereas the other half consists of confusions with STRONGANGER and NONANGER. The performance of the STRONGANGER class ranges with $f1 = 0.64$ between NONANGER and SLIGHTANGER. While only half of the STRONGANGER samples could be identified ($\text{recall}_{\text{STRONGANGER}}$), they have been nearly always correct ($\text{precision}_{\text{STRONGANGER}}$). Since it can be assumed that not all test persons acted with high diligence when voting, we take a closer look at the best performing raters 3 and 16. Both subjects yield high values for both, NONANGER utterances with $F_{\text{NONANGER}} = 0.78$ and 0.77 as well as for angry utterances with $F_{\text{STRONGANGER}} = 0.77$ and 0.81. The annoyed class here likewise seemed to pose major problems when classifying with $F_{\text{SLIGHTANGER}} = 0.61$ and 0.55. Detailed results are listed in Table B.2.

4.3.3 Speech-Based Emotion Recognition

In the next section we assess the performance of the paralinguistic and linguistic emotion recognition approaches. Both use information from the speech sample, i.e., an isolated spoken utterance.

Paralinguistic Emotion Recognition

The isolated performance of automatic emotion recognition solely based on paralinguistic characteristics (i.e. acoustic and prosodic properties) is assessed in the following section. Initially, we seek for the best-performing discriminative algorithm that is able to distinguish between the emotion classes. For simplicity, we perform this evaluation on the state of the art paralinguistic model P_{std}, since it contains a feature space of lower dimension. This assessment will provide us with benchmark values that we achieve with a common paralinguistic model when being applied on real-life data. Afterwards, we direct our attention to the extended paralinguistic model P_{ext}. We apply the best-performing algorithm obtained in the previous assessment and determine the improvement that emerges from the more comprehensive modeling of the dynamics. Due to the considerable dimensionality of the feature vector representation, we remove irrelevant features during a feature selection process. Our further interest lies in the identification of relevant feature groups and isolated features contributing to the task of classifying real-life emotions. The portability of the paralinguistic model between corpora of different domains and languages is assessed and discussed.

State of the Art Paralinguistic Modeling

The automatic classification of emotions is first assessed on the paralinguistic model representations. First, a number of classifiers is applied on the task allowing a distinction of the best-fitting algorithm.

The static feature vector representing the acoustic and prosodic characteristics of an utterance can be classified by a large number of discriminative classifiers, all of which have their specific advantages and disadvantages. In general, there is no commonly superior classifier that outperforms on all classification tasks. Which algorithm fits best on the specific task of classifying emotions in real-life data, is analyzed in the following evaluation process. We deliberately refer to the related literature for a description of the algorithms, since an explanation would go beyond the scope of this work. A good overview on the described algorithms can be found in Witten and Frank (2005) and more formally in Duda et al. (2001). The analysis is restricted on popular base classifiers and common ensemble methods that combine several subclassifiers. The base classifiers are

- an MLP (Duda et al. 2001),
- an SVM (Platt 1999),
- kNN (Duda et al. 2001),
- the rule learner RIPPER (Cohen and Singer 1999),

and the ensemble methods

- AdaBoost (Freund and Schapire 1995) with an SVM,
- Bagging (Breiman 1996) with an SVM,
- Voting (Kuncheva 2004)S over the base classifiers MLP, SVM, kNN, and RIPPER.

The algorithms have been applied on the training and test set of SC-BAC, MOB and LG-FIELD database and evaluated with tenfold cross validation with linear sampling. Results are listed in Table 4.5.

Table 4.5 Classifier performance on static modeling of SC-BAC, MOB and LG-FIELD

Model	Classifier	Database	Class	Recall (%)	Precision (%)	F-measure	f1-measure
P_{std}	SVM	SC-BAC	NonAnger	84.0	78.5	0.81	**0.76**
			Anger	66.7	74.2	0.70	
		MOB	NonAnger	83.9	74.7	0.79	**0.72**
			Anger	59.6	72.2	0.65	
		LG-FIELD	NonAnger	78.4	74.0	0.65	**0.70**
			Anger	67.9	72.9	0.75	
P_{std}	RIPPER	SC-BAC	NonAnger	80.1	76.8	0.78	0.72
			Anger	63.8	69.1	0.66	
		MOB	NonAnger	79.8	71.7	0.74	0.66
			Anger	54.4	64.3	0.57	
		LG-FIELD	NonAnger	73.0	71.9	0.72	0.68
			Anger	63.8	67.4	0.65	
P_{std}	MLP	SC-BAC	NonAnger	75.6	76.0	0.76	0.70
			Anger	65.4	65.3	0.65	
		MOB	NonAnger	75.9	73.6	0.74	0.67
			Anger	61.2	63.2	0.61	
		LG-FIELD	NonAnger	69.9	69.7	0.69	0.64
			Anger	60.1	60.8	0.60	
P_{std}	kNN	SC-BAC	NonAnger	74.3	65.3	0.69	0.58
			Anger	43.0	53.7	0.47	
		MOB	NonAnger	68.5	60.5	0.64	0.52
			Anger	36.6	44.8	0.40	

(continued)

Table 4.5 (continued)

Model	Classifier	Database	Class	Recall (%)	Precision (%)	*F*-measure	f1-measure
		LG-FIELD	NONANGER	56.5	54.3	0.55	0.52
			ANGER	48.0	50.2	0.49	
P_{std}	AdaBoost SVM	SC-BAC	NONANGER	84.3	78.1	0.81	0.75
			ANGER	65.6	74.6	0.69	
		MOB	NONANGER	86.7	74.0	0.79	0.71
			ANGER	56.2	73.7	0.62	
		LG-FIELD	NONANGER	76.3	75.1	0.75	0.72
			ANGER	68.0	73.1	0.69	
P_{std}	Bagging SVM	SC-BAC	NONANGER	84.7	78.0	0.81	0.75
			ANGER	65.4	75.0	0.70	
		MOB	NONANGER	87.3	73.7	0.79	0.71
			ANGER	55.3	74.1	0.62	
		LG-FIELD	NONANGER	78.1	72.4	0.74	0.70
			ANGER	63.6	71.0	0.66	
P_{std}	Voting over SVM kNN RIPPER MLP	SC-BAC	NONANGER	84.2	77.9	0.81	0.75
			ANGER	65.2	74.4	0.69	
		MOB	NONANGER	86.6	74.1	0.79	0.71
			ANGER	56.5	73.5	0.62	
		LG-FIELD	NONANGER	77.8	72.8	0.74	0.70
				64.4	71.4	0.66	

Upper part isolated classifiers, *lower part* ensemble methods

Table 4.6 Feature groups and performance on the SC-BAC, MOB and AIBO database

Feature group	f1 performance on SC-BAC	f1 performance on MOB	f1 performance on AIBO	Number of features
Pitch	0.73	0.68	0.63	240
Loudness	0.71	0.68	0.67	171
MFCC	0.68	0.69	0.71	612
Spectrals	0.69	0.68	0.64	75
Formants	0.68	0.68	0.65	180
Intensity	0.74	0.69	0.69	171
Other	0.67	0.56	0.62	10

The strongest base classifier turns out to be the SVM with polynomial kernel and a degree of one ($C = 0.01$). Following at some distance, the RIPPER algorithm obtains satisfying results and outperforms the MLP. Best results for the MLP could in average be obtained with a sigmoid activation function and one hidden layer with 27 neurons. kNN yields only unsatisfying values with best results obtained using $k = 5$ neighbors for SC-BAC and LG-FIELD and $k = 7$ for MOB. The ensemble methods bagging and AdaBoost have been applied on the most promising classifier as obtained from the evaluation, the SVM. Although applying the bagging approach on SVMs may yield significantly better results (Kim et al. 2002), this is not the case on the described emotion classification problem. Boosting the SVM with AdaBoost yields slightly higher performance values, which, however, turned out not to be statistically significant. In contrast to Morrison et al. (2007), who obtained slightly better performance scores through voting, we cannot report an improvement by voting over SVM, RIPPER, MLP, and kNN predictions. The results confirm the findings from Schuller (2006), who evaluated classification algorithms with acoustic feature vectors of emotion recognition tasks. Obviously basic SVMs outperform both, other base classifiers and ensemble methods. Another important aspect of the presented task is the class-wise performance. Here, the SVM achieved likewise better scores for both classes, ANGER and NONANGER, for virtually all data sets. The results imply the use of SVMs as general classifiers for acoustic modeling and suggest going without ensemble methods.

Extended Paralinguistic Modeling

Under employment of the extended paralinguistic model we determine the relevant feature groups for classifying anger. Table 4.6 shows the different audio descriptors and obtained f1 performance as well as the number of features belonging to a group. It should be noted that the different number of features can take bias on the performance comparison.

The intensity of the speech signal seems to be a good discriminating factor outperforming most other feature groups in the datasets. It should be noted that the pitch features in the English SC-BAC corpus achieves comparatively high scores, while they underperform in the German MOB and AIBO databases.

Further insight can be gained when examining individual feature performance, e.g., produced by a feature ranking scheme. We create an entropy-based information-gain-ratio (IGR) ranking (Witten and Frank 2005) of all 1,477 individual features. Given a single feature and the observed values it holds, the IGR generally estimates the reduction of uncertainty about a class distribution given the conditional entropy of observations. Analyzing the ranked features we see that those features derived from the spectral domain, e.g., MFCC and loudness seem to be most promising for all three databases. They account for more than 50 % of all features. However, MFCCs occur more frequently among the top-ranked features when operating on the MOB database, while loudness features are more frequently among the top ranks when operating on the SC-BAC database. On the AIBO database, too, loudness is highly ranked more frequently. Pitch features account for approximately 25 % of the top sets when trained on the SC-BAC database, while the number is as small as about 10 % when trained on the MOB and AIBO top sets.

The advantage of the extended paralinguistic feature representation is the detailed modeling of the acoustic and prosodic contours. The vector, however, entails a large number of irrelevant and highly correlating features. In machine learning, a high dimensionality of the feature space coupled with a low number of training data can cause the "Curse of Dimensionality", also known as the Hughes Effect (Hughes 1968). It propagates that too many (potentially irrelevant) features might harm the classifier's performance. In order to reduce the number of features and to limit the risk of causing the Hughes effect we perform feature selection to lower the number of irrelevant parameters. Looking for the optimal feature set we incrementally admit a greater number of top-ranked features for classification. The optimization set is employed to determine the optimal feature set size for each data set. Figure 4.6 shows the f1 development as the feature space increases.

The saw-like shape of the graphs indicates a non-optimal ranking, i.e., some inclusions seem to harm the performances. This is mainly due to heuristic IGR estimation. Regarding the magnitude of the jitter we note that it is only about 0.02, which after all proves a generally reasonable ranking. The filter seems to predict best for the MOB database, where the observed jitter is very low. Regarding efficiency, we note that including only 120 features for the MOB, and 100 features for the AIBO database results in a loss for f1 of only about 0.01. To avoid overestimation we apply 10-fold stratified cross-validation for any classification steps. The optimal feature-set size is 220 for the German MOB and AIBO databases and 80 for the English database.

Performance Discussion

Table 4.7 shows the scores obtained when the train/test set is applied using the optimized feature sets with an SVM. Again, 10-fold cross-validation with linear sampling has been applied.

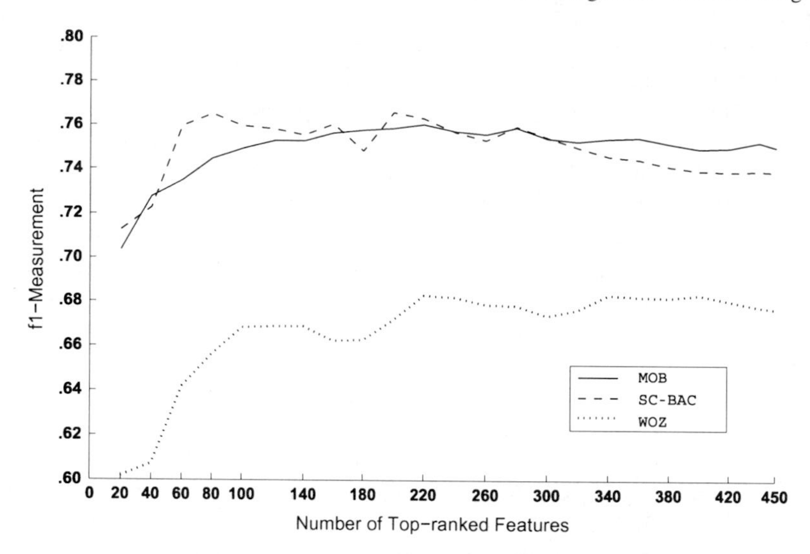

Fig. 4.6 Determination of optimal feature set size using an SVM and an IGR ranking scheme

Table 4.7 Classification results for extended paralinguistic modeling

Model	Database	Class	Recall (%)	Precision (%)	F-measure	f1-measure
P_{ext}	SC-BAC	NONANGER	83.6	80.1	0.82	0.77 (+0.01)
		ANGER	68.8	74.0	0.71	
P_{ext}	MOB	NONANGER	86.3	79.0	0.82	0.76 (+0.04)
		ANGER	64.7	75.8	0.69	
P_{ext}	AIBO	NONANGER	85.1	77.0	0.81	0.68
		ANGER	50.0	63.6	0.56	

Comparing the scores of the IVR databases SC-BAC and MOB obtained from P_{ext} with the scores obtained from P_{std} (cf. scores for SVM in Table 4.5), we observe an increase for both databases. The detailed modeling of the acoustic contours and the new features improve classification particularly for the MOB database. Improvements compared to basic paralinguistic modeling amount to 0.04 points absolute for MOB and 0.01 points absolute for SC-BAC. It can be assumed that the duration of the utterances plays a decisive role on the performance gain. While the previous basic paralinguistic model seems to sufficiently capture the acoustic and prosodic properties of the *short* speech signals from SC-BAC, it fails on *longer* utterances such as in the MOB dataset. It should be noted that the average utterance length in SC-BAC amounts to 0.8 s. In MOB, utterances last more than twice as long with 1.8 s. Longer lasting utterances are thus better modeled.

The F-measures of the ANGER classes are approximately 0.70, while the F-measures of NONANGER classes exceed this figure by 0.10 absolute. Acoustic modeling of the ANGER class of the AIBO corpus seems problematic. The

Table 4.8 Classification results using linguistic information (BOW, TF, TF-IDF, EmoSal, and SRI) when applying 10-fold cross-validation and an SVM with linear kernel

Model		SC-BAC	MOB	AIBO
BOW	f1 $\mid F_{\text{ANGER}} \mid F_{\text{NONANGER}}$	0.47|0.21|0.72	0.50|0.40|0.64	0.44|0.15|0.74
TF	f1 $\mid F_{\text{ANGER}} \mid F_{\text{NONANGER}}$	0.47|0.23|0.71	0.51|0.39|0.63	0.47|0.19|0.74
TF-IDF	f1 $\mid F_{\text{ANGER}} \mid F_{\text{NONANGER}}$	0.46|0.20|0.72	0.51|0.37|0.64	0.46|0.13|0.75
EmoSal	f1 $\mid F_{\text{ANGER}} \mid F_{\text{NONANGER}}$	**0.58|0.76|0.40**	**0.64|0.74|0.53**	**0.63|0.78|0.49**
SRI	f1 $\mid F_{\text{ANGER}} \mid F_{\text{NONANGER}}$	0.58|0.66|0.51	0.63|0.69|0.56	0.68|0.74|0.62

Evaluated databases are SC-BAC, MOB, and AIBO

models were not able to capture the emotion-related information needed for classification. As a hypothetic explanation, this could be due to the mapping of the classes *touchy*, *angry*, *reprimanding*, and *emphatic* into the single cover class of ANGER. Although the inter-labeler agreement does not indicate human differences in perception between the German databases MOB and AIBO, the acoustic models might been blurred by the process of sub-summation of classes.

Linguistic Modeling

To assess the predictive value of linguistic modeling on real-life speech corpora, we classify the vector space models with SVMs using a linear kernel. Table 4.8 shows the classification results of the linguistic features for SC-BAC, MOB, and AIBO obtained by 10-fold cross-validation. For SRI and Emotional Salience we apply an out-of-vocabulary (*OOV*) handling when the individual test splits of the cross-validation folds contain words of unknown SRI or Emotional Salience. If no SRI or Emotional Salience value exists for a word, its contribution to the turn summation is skipped. Turns completely consisting of OOV words are classified as belonging to the majority class. For SC-BAC we determined an average OOV-rate of 5 %, for MOB and AIBO respectively 5 %, both of which are relatively low figures.

Impact of Corpus-Related Influence Factors on SRI and EmoSal Performance

SRI and Emotional Salience are dependent on word posteriors, which in turn depend on word counts. Assuming that different conditions in the corpora influence the performance we examine the impact of the different corpora designs on the classification task. Analyzing the impact of differences in number of turns, total number of words, and vocabulary size we generate 10 randomly chosen subsets retaining original class distributions. Original sizes and targeted, sub-sampled sizes are given in Table 4.3. Estimating averages from the random sets we match the conditions between the databases, e.g. by sub sampling the SC-BAC corpus to match the vocabulary size of MOB. Each randomly generated subset is further processed by 5-fold cross-validation to calculate SRI and Emotional Salience performances. It can be

Table 4.9 SRI and Emotional Salience of phrases

	SC-BAC	MOB	AIBO
f1 SRI all words	0.58	0.62	0.67
f1 Emotional Salience all words	0.57	0.62	0.63
f1 SRI minimum 4 words	0.57	0.62	0.67
f1 Emotional Salience minimum 4 words	0.56	0.61	0.64

reported that the performance scores on none of the sub-sampled versions of the original corpora considerably differed from the original conditions. Hence, SRI and Emotional Salience seem relative robust with regard to varying vocabulary sizes and class distributions.

Perplexity. When matching the dictionary sizes between MOB and AIBO corpora, the MOB corpus shrinks to 58 % of its original number of turns while the perplexity almost stays constant, i.e. it merely shrinks from 233 to 223. The perplexity expresses the average word branching factor of a language model, cf. Rabiner (1990); Pfister and Kaufmann (2008). A large perplexity value indicates a more diverse language model whereas a lower value is usually obtained for more specialized language models. As can be seen from Table 4.3 the databases essentially differ in perplexity, which in other words expresses the amount of confusion when choosing uniformly and independently among all words. Unfortunately, sub-sampling the database to match the perplexity condition does not yield sufficient volumes of data for training. A preliminary classification experiment on the basis of these sets showed an intolerably high standard deviation among the results from cross-validation splits.

Emotional Salience for Sentences. Expanding the basic modeling unit from separated words to phrases we include contextual word information by calculating Emotional Salience of phrases, see e.g. Metze et al. (2009). Because of a low average word-per-turn rate, we apply phrase modeling including two consecutive words only. The resulting vocabularies comprise 4,053 entries for AIBO, 4,973 for MOB, and 880 entries for the SC-BAC database. Table 4.9 shows the results when all phrases are taken into account, regardless of their frequency of occurrence in the sets. In order to obtain more robust estimates we set a frequency threshold. Table 4.9 also shows the scores when admitting only phrases (and words) that occur more than 4 times.

The results show that the inclusion of word context does not improve emotion recognition scores on the analyzed corpora. Also, the use of n-gram models directly has been proposed (Steidl 2009; Shafran 2005) as well as other linguistic features, e.g. part-of-speech (*POS*) or higher semantics representations as reported on by Schuller et al. (2009). In the literature, none of the respective techniques resulted in significant improvements for the respective corpora. It should be noted that all experimental databases in these works were comparable to those employed in this work and are of low average word-per-turn rate as well. Hence, an interpretation of the effect of linguistic context modeling to emotion classification has to be deferred to future experiments including databases with longer utterances. However, Steidl

(2009) reports on different word clustering techniques that, after all, seem to be most promising for databases with very short utterances. Since these methods need additional labeling, the respective features will not be available for most applications.

Hand-Transcripts Versus ASR Transcripts

The previous evaluations were conducted using manual transcripts from human transcriptionists, which does not reflect the conditions required for an automatic linguistic emotion recognition system. In the following we thus direct our attention on the impact of word hypotheses quality, as obtained from the ASR module. Metze et al. (2009); Polzehl et al. (2009b) as well as Schuller et al. (2009) report Word Error Rates (*WER*) of less than 20 % and more than 30 %, respectively. At the same time, the impact of the errors on anger recognition performance remains limited.

In the next evaluation experiment we therefore focus on the dependency of emotion recognition from text upon automatic transcription quality. We simulate ASR quality by systematically varying the decoding beam of a speech recognizer. Narrowing the beam of the speech recognizer resulted in a monotonously rising word error rate. Analyzing the impact on emotion classification it can be observed that anger classification stays roughly constant as long as the word accuracy (WA) does not fall below a certain threshold. Using Emotional Salience models this threshold was found to be around 40 % WA for the German AIBO database, before the recall curve considerably degrades. Comparing the best WA of 82 % with a downgraded WA of 40 %, we observe only a small decrease of roughly 5 % in emotion recognition. Once the WA falls below 40 %, the emotion recognition performance visibly degrades. As an interesting result, word errors are not strongly harming the emotion recognition system as long as the errors occur consistently: wrongly recognized words are likewise correlated to emotion classes when building the Emotional Salience model, and when the same error occurs in decoding, the erroneous token nevertheless link to the respective emotion class.

Focusing on the impact of training–testing mismatch we analyzed the performance difference between *matched* training, i.e. the model has been trained and tested with ASR transcripts, and *mismatched* training, i.e. the model has been trained with human transcripts and tested with ASR transcripts. Mismatched training outperformed matched training merely by about 2 %. Hence it seems not essential to have high-effort transcription for training in order to achieve good results when testing. As long as ASR errors occur systematically the emotion models will capture the class relevant information, even given erroneous word hypotheses.

Performance Discussion

Taking a closer look at the performance values in Table 4.8 it can be seen that features originally intended for document classification and text mining like BOW, TF, and TF-IDF fail for emotion recognition on all three employed data sets. In contrast to this finding Schuller (2006) reports satisfying results for TF-IDF on a 7 class recognition task, which might be due to the considerably longer phrase length (20 compared to 1–2 in the employed data sets).

Looking at the F-measures of the individual classes we can see that too many test samples were classified as NONANGER. Another drawback of the BOW-related models is that unknown test words in the cross-validation splits cannot be assigned to the majority class. All feature models stay close to (or do not even reach) the majority baseline of $f1 = 0.50$.

While the F-measure of the NONANGER class degrades when using SRI and Emotional Salience, they achieve better F-scores for spotting the ANGER class. Due to the weighting factor, Emotional Salience seems more robust to imbalanced class distributions while SRI is more influenced by the majority class. However, it should be noted that the absolute performance of SRI and Emotional Salience does not exceed the prior class probabilities, i.e. the baseline of $f1 = 0.50$, considerably. For further experiments we retain the most promising features, i.e. SRI and Emotional Salience.

4.3.4 Dialog-Based Emotion Recognition

The approaches that are specifically designed to support emotion recognition in spoken human–machine interaction and which use information adhering to a dialog are assessed in the following.

Interaction and Context-Related Modeling

The predictive efficiency of interaction- and context-related information is assessed using the SC-BAC and LG-FIELD databases as MOB and AIBO are not endowed with interaction parameters. For this evaluation we omit the feature selection due to the comparably low number of interaction parameters. For both corpora, however, we first assess the contribution of the single interaction parameters using an IGR ranking that has also been used for feature selection in the extended paralinguistic model, see Sect. 4.3.3. The ranking gives some indication of the relevance of isolated parameters and is depicted in Table 4.10. Ranging among the top 15 most relevant parameters the following features occur in both databases: %ASRREJECTIONS, #BARGE- INS, #REPROMPTS, #SYSTEMTURNS, %TIME-OUTS_ASRREJ, #TIMEOUTS_ASRREJ, %TIME- OUTPROMPTS, TURNNUMBER and UTD. The ASR rejection rate (%TIMEOUTS_ASRREJ) and the rate of time-outs (%TIMEOUTS_ASRREJ) are relevant discriminating parameters for anger detection. Apparently, a large number of such rejections and time-outs appear to displease users as both events cause the DM to re-prompt the question. At this point, the system reveals that it has not correctly understood the user, which in return abets anger. The duration of the user response (UTD) shows to be a further discriminator for anger

Table 4.10 Top 15 interaction parameters for discriminating between ANGER and NONANGER obtained by IGR ranking on subsets employed for anger recognition from SC-BAC and LG-FIELD

Database	SC-BAC		LG-FIELD	
Rank	Feature	IGR	Feature	IGR
1	**% TIME-OUTPROMPTS**	1.00	% REPROMPTS	1.00
2	# ASRSUCCESS	0.95	**# REPROMPTS**	0.89
3	**# REPROMPTS**	0.90	# USERTURNS	0.75
4	**TURNNUMBER**	0.88	# ASRREJECTIONS	0.66
5	% HELPREQUESTS	0.88	**TURNNUMBER**	0.62
6	% UNEXMO	0.85	**# SYSTEMTURNS**	0.62
7	**UTD**	0.82	% BARGE- INS	0.62
8	**% ASRREJECTIONS**	0.82	**UTD**	0.61
9	**# BARGE-INS**	0.78	# SYSTEMQUESTIONS	0.60
10	**# SYSTEMTURNS**	0.78	# TimeOuts_ASRRej	0.58
11	**% TIMEOUTS_ASRREJ**	0.76	DD	0.57
12	% OPERATORREQUESTS	0.74	# BARGE- INS	0.57
13	% ASRSUCCESS	0.73	**% ASRREJECTIONS**	0.55
14	WPUT	0.71	**% TIMEOUTS_ASRREJ**	0.43
15	**# TIMEOUTS_ASRREJ**	0.71	**% TIME-OUTPROMPTS**	0.38

Parameters occurring in both top-ranks are marked in bold-face

as a longer user response, in contrast to short commands, implies dissatisfaction. Users frequently interrupting the system prompt (#BARGE- INS) furthermore seem to get more frequently annoyed as a high absolute number of barge-ins implies anger. The immediate barge-in behavior of the user has no influence due to the absence of context-related barge-in parameters.

10-fold cross-validation on both datasets is carried out for assessing the performance of interaction- and context-related models. Highlighting the generalization of the model we furthermore assess the cross-corpus performance. Therefore, nominal-valued interaction parameters have been excluded as the values cannot be reasonably mapped between the two domains. Training has respectively been conducted with the SC-BAC set and evaluated with the LG-FIELD set and vice versa. Prior to training and testing each data set has been pre-processed by a z-score normalization function to adapt the different scales of the respective interaction parameters in both corpora.

Performance Discussion

Performance scores for interaction- and context-related modeling are depicted in Table 4.11. Given that neither paralinguistic nor linguistic information sources are included in the model, a remarkable f1-measure of 0.64 is obtained for both data sets. Comparing these results to paralinguistic modeling for SC-BAC,[2] the context-

[2] see SC-BAC performance in Tables 4.5 and 4.7

Table 4.11 Performance of the interaction- and context-related modeling.

Model	Database	Class	Recall (%)	Precision (%)	F-measure	f1-measure
I and C	SC-BAC	NONANGER	83.5	68.6	0.75	0.64
		ANGER	44.5	65.0	0.53	
I and C	LG-FIELD	NONANGER	75.1	68.0	0.70	0.64
		ANGER	58.8	67.0	0.58	

Evaluated on 10-fold cross-validation with linear sampling and an SVM with linear kernel

Table 4.12 Performance of the interaction- and context-related modeling when *cross-corpora evaluation* is applied using an SVM with linear kernel

Train database	Test database	Class	Recall (%)	Precision (%)	F-Measure	f1-measure
LG-FIELD	SC-BAC	NONANGER	77.29	66.00	0.71	0.60
		ANGER	42.24	56.18	0.48	
SC-BAC	LG-FIELD	NONANGER	87.34	58.80	0.70	0.55
		ANGER	28.59	65.93	0.40	

and interaction-related model performs weaker. However, in comparison to linguistic modeling it outperforms all linguistic models BOW, TF, TF-IDF, EmoSal, and SRI.[3]

Considering the class-wise F-measures we observe strong performance differences for predicting the two classes. While the model achieves satisfying F-measures for NONANGER for both, corpus-internal and cross-corpus evaluation, it seems less suitable for predicting the ANGER class. Here, EmoSal, SRI as well as both paralinguistic models show stronger results.

As visible from cross-corpus evaluation scores in Table 4.12, interaction- and context-related modeling generalizes to a certain extent. SC-BAC reaches an f1-measure of 0.60, which depicts only 0.04 points less than the corpus-internal evaluation (cf. Table 4.11). Training on and thus 0.09 points below the score obtained by corpus-internal evaluation. Corpus- and domain-dependent parameter scales account for the lower performance of this cross-corpus evaluation in comparison to corpus-internal validation. The results show that obviously domain- and corpus-independent properties of an interaction exist indicating whether a user is likely to show anger.

Emotional History Modeling

To model the history of the user's emotional state we proceed as described in Sect. 4.2.2 and map the described approach to a binary anger detection task. Let $\varepsilon(t) \in \{A, N\}$ be an emotional state of a user at dialog turn t, where A depicts the emotion class ANGER and N the emotion class NONANGER. Further let $\Omega = \{\varepsilon(t_1), \ldots, \varepsilon(t_{n-2}), \varepsilon(t_{n-1})\}$ be an observation sequence of emotional states ε_i occurring in the respective user turns t_i prior to the currently observed user turn t_n. Then

[3] see SC-BAC performance in Tables 4.8 and 4.11

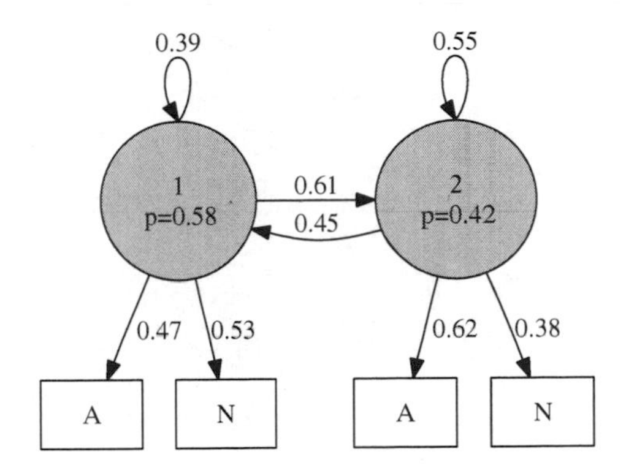

Fig. 4.7 HMM architecture including priors, transition, and emission probabilities after training with data from SC-BAC. The depicted HMM contains the probabilities for observing an angry turn

we can discriminate between two types of sequences, namely Ω_{ANGER}, observations leading to an angry user turn, and Ω_{NONANGER}, leading to a non-angry user turn. The probability of the emotional state $\varepsilon(t_n)$ belonging to the class ANGER or the class NONANGER can be determined by two HMMs M_1 and M_2, where

$$M_1 = (\mathcal{S}, \mathcal{A}, \pi, \Omega_{\text{ANGER}}, \mathcal{B}) \tag{4.10}$$

and

$$M_2 = (\mathcal{S}, \mathcal{A}, \pi, \Omega_{\text{NONANGER}}, \mathcal{B}) \tag{4.11}$$

with $\mathcal{A}$ being the respective state transition probabilities, $\mathcal{B}$ the observation symbol probability distribution, and π the initial state distribution obtained by training.

The employed model is endowed with the two hidden states $\mathcal{S} = \{S_1\ S_2\}$, see Fig. 4.7. Each time we transit from one state to the other, or decide to stay in the same state, one of the two observation symbols "A" or "N" is emitted.

For each of the two possible emotion classes, ANGER and NONANGER, a separate HMM is trained with observation sequences, respectively leading to anger or nonanger. During the training process the emission and transition probabilities in each of the two HMMs are adapted. Based on the prior emotional states of the user the production probability of the current emotional state can be determined. Both HMMs evaluate the observation sequence using the respective transition and emission probabilities. The generated likelihood expresses how likely the specific HMM generated this observation.

The current turn t_i belongs to the class ANGER if the condition

$$P(\Omega|M_1) > P(\Omega|M_2) \tag{4.12}$$

is satisfied.

To assess the performance of the model we employ the SC-BAC and MOB databases. The evolvement of anger in both employed subsets is depicted in Fig. 4.8.

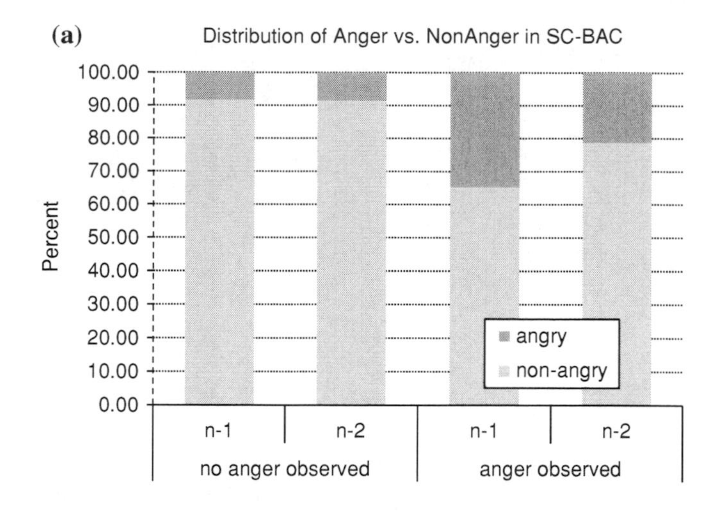

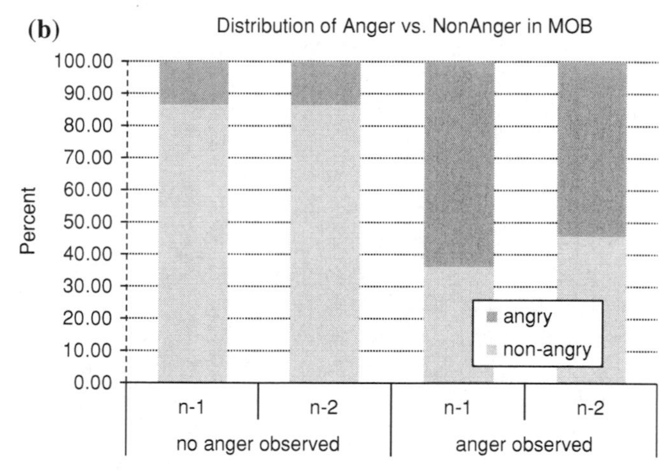

Fig. 4.8 Anger state in the two previous user turns of the currently considered one: when observing an angry user turn the likelihood of observing another one is 34.7 % in SC-BAC (**a**) and 63.8 % in MOB (**b**)

In SC-BAC, the likelihood of the previous turn of an ANGER user turn being also ANGER can be reported with 34.8 %. For the pre-predecessor this probability is still 21.3 %. For MOB, the dependency between ANGER turns is even stronger: the probability of a predecessor of an ANGER user turn being also ANGER is 63.8 %, for the pre-predecessor still 54.3 %.

For evaluation, we apply 10-fold cross-validation and evaluate the model for each of the two corpora. The observation sequences representing the anger history are based on manually annotated labels. The results are depicted in Table 4.13.

Corpus-related differences are clearly visible when looking at the obtained performance scores. As can be seen from the results, the model performance depends

Table 4.13 Performance of the history modeling for SC-BAC and MOB, evaluated with an HMM trained on entire corpus

Model	Database	Class	Recall (%)	Precision (%)	*F*-measure	f1-measure
H	SC-BAC	NONANGER	79.48	64.60	0.71	0.58
		ANGER	36.82	55.30	0.44	
H	MOB	NONANGER	83.33	77.04	0.80	0.74
		ANGER	62.72	71.48	0.67	

on the occurrence of anger in a data set. With SC-BAC exhibiting a lower inter-turn dependency of anger, i.e. anger occurs more isolatedly, the performance is likewise worse compared to MOB. In general, the prediction of NONANGER in both corpora yields better performance scores compared to ANGER, the overall f1-measure for MOB clearly comes close to the performance of the extended paralinguistic should be noted that a slight decrease in performance may be expected when using classifier predictions instead of manually annotated labels. Furthermore, the model performance presumably degrades when a larger number of classes is introduced. Due to the lack of appropriate data, this assessment has to be postponed to future work.

4.3.5 Fusion Strategy

Different strategies exist for bringing together the models to reach a final decision in classification, see Fig. 4.9. Early fusion (also denoted as feature level fusion) joins the feature spaces prior to classification by constructing a large feature vector. A limiting factor for this approach is the Hughes effect, i.e. the curse of dimensionality (Hughes 1968), since the feature space might grow considerably. This again may be limited by a joint feature selection process, e.g. with IGR or Genetic Algorithms (Duda et al. 2001). Late fusion (also denoted as decision level fusion) relies on the prediction of model-specific classifiers joining the hypotheses, confidence scores, or similar information. The final decision hereby may be achieved by simply averaging the confidences of the isolated classifiers (Lee and Narayanan 2005; López-Cózar et al. 2008) and deciding for the class with maximum confidence. Simply averaging over isolated classifier's confidence scores would yield unsatisfying results with our data sets as some sub-classifiers show inferior results than others. This might thereby degrade the entire performance. Alternatively, a decision may be achieved by creating a final feature vector that, again, is subject to another classifier (see Fig. 4.9b).

Paralinguistic and Linguistic Fusion

At first, we analyze the performance of the extended paralinguistic model jointly with the linguistic models. As previous experiments using early fusion techniques

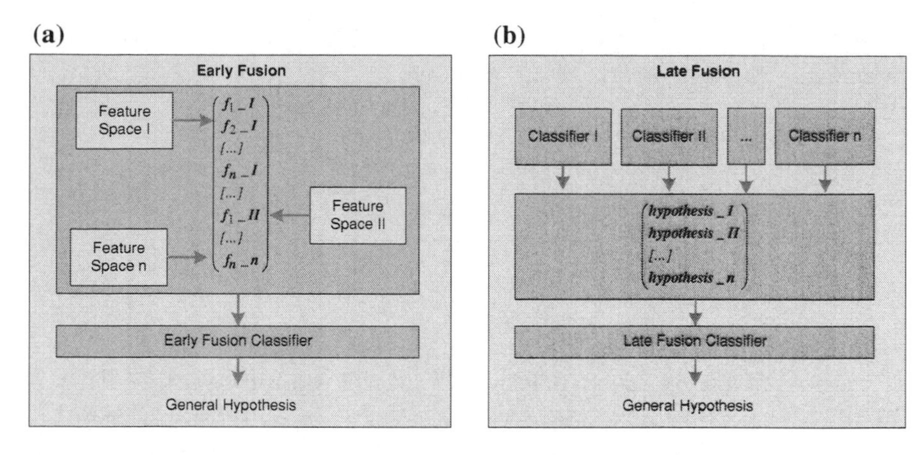

Fig. 4.9 Different fusion techniques: in early fusion, **a**, feature spaces of the models are merged before they are subject to classification; in late fusion, see **b**, classification takes place for each respective model and the class hypotheses, class-wise confidence scores or further output are subject to a final classifier

by merging feature spaces show inferior results (Polzehl et al. 2011), which might be attributed to the high dimensionality of the joint feature vector, we combine the predictions obtained from the respective models by late fusion. The single classifiers provide hypotheses, or class-wise confidence scores that are merged in form of a feature vector, which is used for training the final classifier, see Fig. 4.9.

The final feature space consists of eight features, four from each acoustic and linguistic system. Both systems contribute a predicted class. The acoustic system further generates scores by logistic regression analysis for each prediction. The linguistic system contributes the estimated SRI scores directly. On the basis of these scores we compute normalized confidence estimates for both classifiers by computing the rank for a score in its population and re-normalizing this to a range of $[0, 1]$. In our case, the normalized rank corresponds to the discrete probability distribution. In other words, we estimate the amount of confidence in a prediction to belong to a class by considering likelihoods of predictions of higher and lower values for that class.

Classifying the fused hypothesis vector we assess the performance of SVM with a linear kernel, an SVM with radial basis function kernel and an MLP. In the MLP experiments we use back-propagation training in maximal 500 epochs. A validation set size of 10 % is used for early training termination. The nodes of the three layers are connected by sigmoid activation functions; the middle layer comprises 4 nodes. We further extent the SVM classifier to nonlinear mapping by transforming data using an RBF kernel function. We determine the optimal settings for the algorithm's complexity parameter and the kernel width by 10-fold cross-validation on the train set in a grid-search manner.

Table 4.14 shows the final classification scores obtained by 10-fold cross-validation.

Table 4.14 Classification performance after fusion of P_{ext} and L models on decision level in comparison to the paralinguistic baseline (P_{ext} only)

Type	Models	Baseline	Database	Classifier	F-measure Anger	F-measure NonAnger	f1
Late	$P_{ext} + L$	P_{ext}	SC-BAC	SVM-linear	0.70	0.82	0.76 (−0.01)
				SVM-RBF	0.73	0.82	**0.78 (+0.01)**
				MLP	0.73	0.81	0.77 (±0.0)
Late	$P_{ext} + L$	P_{ext}	MOB	SVM-linear	0.70	0.82	0.76 (±0.0)
				SVM-RBF	0.73	0.83	**0.78 (+0.02)**
				MLP	0.72	0.82	0.77 (+0.01)
Late	$P_{ext} + L$	P_{ext}	AIBO	SVM-linear	0.55	0.81	0.68 (±0.0)
				SVM-RBF	0.57	0.83	**0.70 (+0.02)**
				MLP	0.57	0.83	0.70 (+0.02)

SVM-RBF yields the best scores

As expected, the fusion of both types of information at the decision level generates slight improvements for all corpora. Looking at the magnitude of the difference, little improvement could be expected due to inferior performance of the linguistic classification. After all, the final scores resemble the scores obtained from the acoustic classification. We note that for the approximately 33/66 split of class distributions for the AIBO corpus, constant classification into the majority class would result in approximately 0.40 f1. For the IVR databases constant majority class voting would result in 0.38 f1. Similarly, to the result of the acoustic classification, the low F-measures for the ANGER class models for the AIBO database turns out to be problematic. Another expected result is that nonlinear classification yields better results than linear classification. Furthermore, in our experiments SVM-RBF slightly outperforms MLP fusion.

Paralinguistic and Interaction- and Context-Fusion

As late fusion between the paralinguistic classifier and the interaction- and context-related classifier showed inferior results compared to the isolated classifiers, we fused both feature spaces prior to classification in an early manner. The results in Table 4.15 show performance gains when paralinguistic and interaction- and context-related features are employed. This particularly holds true for LG-FIELD.

0.01 and even 0.07 points can be gained for SC-BAC and LG-FIELD respectively when interaction- and context-related information is applied. The joint system thereby reaches the same performance for SC-BAC as P_{ext} model. However, it achieves this score with much less computational costs as the employed interaction parameters can be derived at virtually no costs from interaction logs. It seems advisable to employ the basic paralinguistic model jointly with interaction-related information for emotion recognition in SDS.

Table 4.15 Early fusion of paralinguistic model P_{std} with interaction- and context-related model I and C (SVM, 10-fold cross-validation with linear sampling)

Type	Model	Baseline	Database	Class	Recall (%)	Precision (%)	F	f1
Early	$P_{\text{std}} + \text{IC}$	P_{std}	SC-BAC	NONANGER	84.83	79.54	0.82	0.77 (+0.01)
				ANGER	68.35	77.65	0.71	
Early	$P_{\text{std}} + \text{IC}$	P_{std}	LG-FIELD	NONANGER	81.19	77.56	0.79	0.77 (+0.07)
				ANGER	72.56	76.78	0.75	

Paralinguistic and History Fusion

Assessing the benefit of history-related models we fuse the paralinguistic model P_{std} with the history model H. Late fusion is applied, since classifiers and feature spaces are not compatible and thus early fusion must be ruled out. The final 4D feature vector consists of the two class-wise confidence-scores from the paralinguistic model P and the two production probabilities P_{ANGER} and P_{NONANGER} generated by the two HMMs. As SVM-RBF has shown superior results in previous experiments, we classify the final vector with SVM-RBF. Results are listed in Table 4.16.

As expected due to the poor isolated performance of the history model in SC-BAC, an improvement cannot be shown when joining SC-BAC's paralinguistic model with the history model. Entirely different are the results for MOB: an improvement of 0.06 points can be reported in comparison to the isolated paralinguistic model.

Performance Discussion

Combining knowledge sources may considerably increase the performance when classifying emotions. Improvements could be reported for all suggested approaches. For fusion, we considered the paralinguistic models as baseline, as they achieved in average the highest scores among the competing models. Thereby any improvement has to be viewed in dependency to the underlying base classifier. While the performance gain achieved by joining linguistic information may appear low in comparison to context- and interaction-related models, it should be kept in mind that the base classifier P_{ext} already achieves higher scores than the P_{std} classifier. No general recommendation can be given when it comes to fusion techniques. Late fusion has not always shown superior results and failed for example for paralinguistic/interaction fusion. If feature dimensionality is low and feature spaces are compatible, an early fusion might achieve better results. Incompatible feature spaces and classifiers should be merged with a late fusion technique. Thereby SVM-RBF has proved superior performance compared to SVM-linear and MLP.

4.3.6 Emotion Recognition in Deployment Scenarios

It has become evident that emotion classifiers for real-life data in general achieve higher scores for the nonanger as for the anger class. This particularly poses a problem for scenarios where an emotion recognizer is deployed in an SDS. Here, the wrong detection of the emotional state and the hereby associated wrong adaptation of the dialog may lead to unpredictable outcomes. We have introduced a cost-sensitive classifier that may favor the recognition of particular emotion classes on the cost of the remaining classes. For assessing the proposed approach, we empirically determine the sensitive cost values for misclassifying angry user turns. $R(i|x)$ has been defined as

Table 4.16 Late fusion of the basic paralinguistic model with the history-related model using SVM-RBF (10-fold cross-validation with linear sampling)

Type	Model	Baseline	Database	Class	Recall (%)	Precision (%)	F-measure	f1-measure
Late	$P_{std} + H$	P_{std}	SC-BAC	NONANGER	84.43	78.39	0.81	0.76 (±0.00)
				ANGER	68.24	74.57	0.70	
Late	$P_{std} + H$	P_{std}	MOB	NONANGER	80.54	82.04	0.81	0.78 (+0.06)
				ANGER	74.96	73.05	0.74	

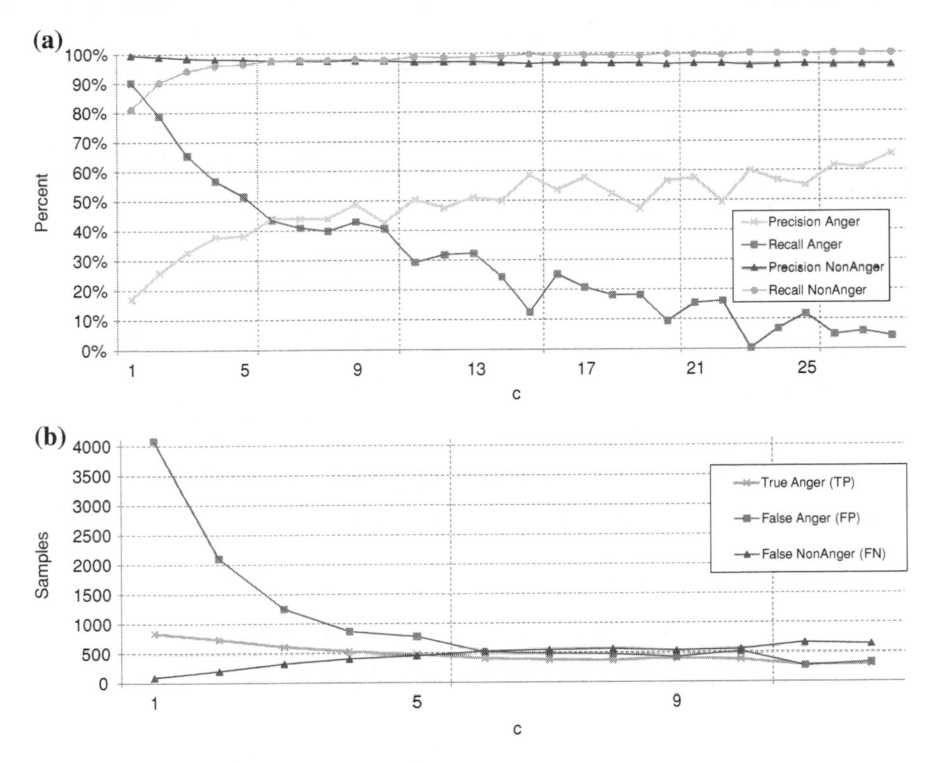

Fig. 4.10 Results of iteratively increasing the cost for misclassifying a non-angry user turn: class-wise precision and recall values (**a**) and number of TP, FP, FN (**b**). For reasons of clarity, the number of TN is not depicted, which slightly increases from 19,000 to 20,000 samples

$$R(i|x) = \sum_j P(j|x)C(i,j). \tag{4.13}$$

We define the cost matrix as follows

$$C(i,j) = \begin{pmatrix} \text{TP} = 0 & \text{FP} = c \\ \text{FN} = 1 & \text{TN} = 0 \end{pmatrix} \tag{4.14}$$

and introduce a dynamic factor for false positives allowing us to penalize stronger the misclassification of non-angry users as angry ones. An optimal value for c is thereby determined in an iterative process empirically, i.e. $c = \{1, 2, \ldots, 30\}$. The SC-BAC data set has been applied for this evaluation. The resulting performance charts are listed in Fig. 4.10.

With increasing penalty the classifier favors for the class non-angry. While the precision for ANGER increases, the recall degrades. Looking at Fig. 4.10 we can see that the optimal cost for misclassifying non-angry turns for SC-BAC is a cost value of 7. It reduces the number of misclassified non-angry turns (FP) by factor 16.9,

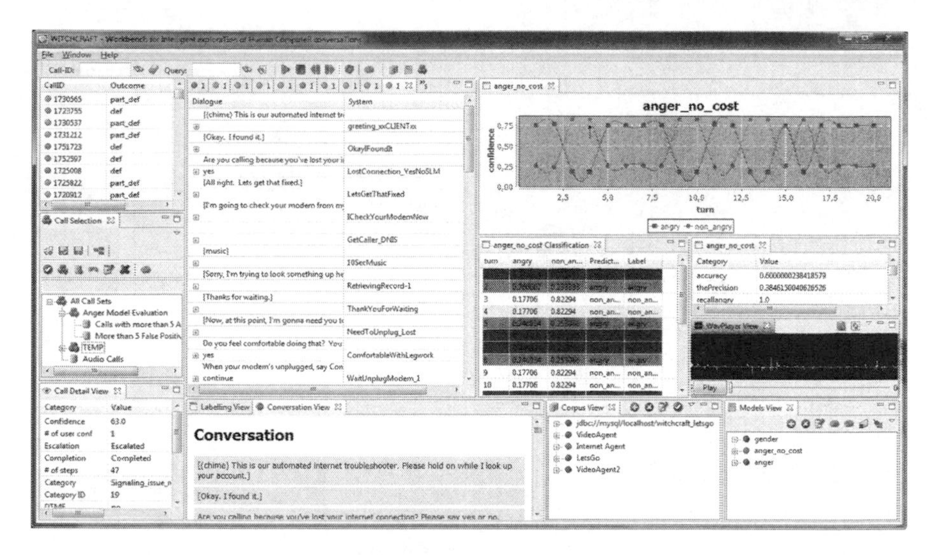

Fig. 4.11 Screenshot of the Witchcraft Workbench in the Analysis Perspective. It consists of two Call Selection Views (*top-left*), a Call Detail view, the central Dialog View enabling navigation within the conversation, Classification Tables, Chart Views, and the audio player. Turns that have been correctly identified as anger (TP) are depicted with *green* background color (here *light grey*), false predictions (FN + FP) are depicted with *red* background color (here *dark grey*)

namely from approximately 5,000 (cost = 1) to 100 while the number of correctly classified turns (TP) is only reduced by factor 3.7 from approximately 400 to 100.

Deployment Simulation in Witchcraft

With the intention of deploying emotion recognition as module for online monitoring the interactions in SDS, one should devote special attention to the effect that a deployment might cause. Joint analyses of the classifier's predictions together with the respective interaction context allow a judgment on the model's suitability for practical applications.

To assess the effect of deploying the emotion recognizer previously introduced, we analyze the classifier within the Witchcraft Workbench. Furthermore, we assess how the adapted cost-sensitive classifier performs in comparison. Witchcraft parses the predictions generated by the classifiers (cf. Sect. 3.3), which have been determined on the exchange level. Next it creates statistics on the dialog level. It counts the number of TP, FP, TN, FN within the dialog and calculates class-wise precision and recall values as well as accuracy, *f*-score, etc. The integrated SQL-based query mechanism allows searching for dialogs fulfilling a specific criterion. By that, dialogs with a high number of false positives, low f1 scores such as for the class ANGER can be spotted.

An example of such a dialog under analysis in the Witchcraft Workbench is depicted in Fig. 4.11.

An auditory analysis of the misclassified samples in Witchcraft turned out that the anger model seems to frequently misclassify distorted and garbage turns that contain higher frequencies and high loudness values similarly as the samples from the ANGER class.

The following evaluation is based on the 1,911 dialogs from SC-BAC. When using the regular classifier Witchcraft identifies 1.196 calls that contain no user anger but at least one mistakenly as angry classified user turn. 313 calls contain even five and more such misclassifications. The adjusted cost-sensitive classifier with the cost value for misclassifying non-angry turns set to "7" produces a different image. The numbers drop to only 134 and four calls respectively. Two representative dialogs are depicted in Fig. 4.12.

The regular classifier generally causes a large number of false alarms. Figure 4.12a contains a dialog from a non-angry user. The regular classifier (left side) misclassifies a high number of non-angry turns as angry ones. The cost-sensitive classifier (right side) ignores the predictions of the underlying classifier and decides for "non-angry". Figure 4.12b contains a dialog from an angry user. The regular classifier predicts nearly all angry turns correctly while the cost-sensitive one behaves more conservatively and misses some of the true angry turns.

Discussion

The impact of deploying anger recognition to an IVR system should not be underestimated. The presented regular classifier generates a very high number of false positives. This first setup does not seem suitable for an IVR system that reacts sensitively on the emotion recognizer's predictions, such as a system relying on anger detection for escalating to operators. Employing the presented cost-sensitive classifier we are able to decrease the number of misclassified non-angry user turns. For this setting, a cost value of 7 turned out to be the best compromise in order to gain low FP values while keeping the TP value comparably high. Notwithstanding these improvements, the false positives for the detection of angry users remain twice as high as the true positives. It should be reminded that we have originally merged the classes SLIGHTANGER and STRONGANGER to one single class ANGER (cf. Sect. 4.3.1), which could explain this behavior. The patterns between very angry and non-angry turns may have been blurred by this process, which causes the classifier to lose performance.

For classifiers that are intended to predict hot anger online it is thus recommendable to adjust the training process and include only those samples that contain hot anger. Further it turns out to be questionable whether the detection of one angry user turn should serve as basis for an escalation to operators. Pattern matching taking into account several user turns would be sensitive to detect users that are in rage.

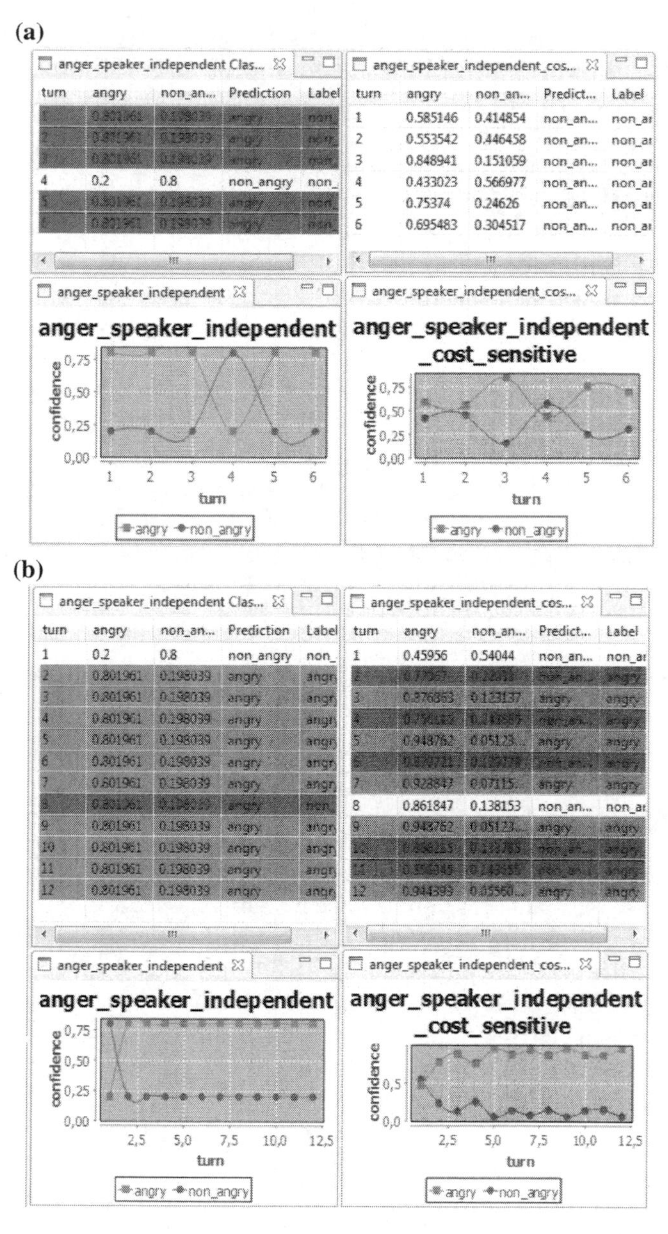

Fig. 4.12 Typical calls from a non-angry user and an angry user predicted from the regular and cost-sensitive learner (respectively *left* and *right sides*). Again, *green* (here *light grey*) turns symbolize correctly spotted angry turns (TP), *red* turns (here *dark grey*) symbolize misclassified turns (FN+FP). **a** Non-angry user. *Left* regular classifier mistakenly classifies a number of NONANGER samples as ANGER. *Right* cost-sensitive classifier identifies NONANGER correctly; **b** Angry user. *Left*: regular classifier mainly spots all ANGER samples correctly. *Right* cost-sensitive classifier identifies some, but not all ANGER samples

4.4 Summary and Discussion

We consider the recognition of emotions as core for online monitoring in SDS as they allow spotting critical dialog situations in spoken dialog interactions. For that reason, we concentrated on the recognition of negative emotions, which may be an indicator for general dissatisfaction and may account for aborting tasks, e.g. for hang-ups in telephone-based systems. Another fundamental reason for focusing on negative emotions was attributed to the fact that other "real" emotions than negative ones do very rarely occur in real-life SDS and may thus be viewed as irrelevant, at the very least for all the three IVR data sets.

In this chapter we have investigated novel techniques for discriminating between emotional states using spoken utterances and presented approaches that use the signal level (i.e. *paralinguistics*) and textual content (i.e. *linguistics*). Particularly intended for the recognition of emotions in SDS we have further introduced novel approaches exploiting knowledge derived from the dialog itself, which are the *interaction- and context-related model* as well as the *history-related model*. Thereby the current emotional state is not only classified based on the actual textual and acoustic characteristic, but also on properties captured from the interaction itself plus the history of the emotion. Our approach takes into account that an emotional state in an SDS does not occur causelessly but is tied to incidents in a dialog, which is modeled through the interaction parameter set. Exceeding the boundary of an isolated user utterance we hereby take into account what "happens" beyond the current exchange. Thereby the joint system achieves performance scores that come close to human performance.

The captured and employed data sets originate in all cases from real-life interactions between users and SDS, i.e. they originate from Human Computer Interaction (HCI) and in all four scenarios the speakers believed to be talking to a machine. The basic comparability of the corpora for an HCI anger task can thus be seen as given. No human–human interaction data has been mixed. We hereby provided benchmark values and pointed out which performance scores in real-life HCI data are realistic. All presented approaches have been evaluated using speaker-independent evaluation techniques through using linear sampling.

Paralinguistic Modeling

With regard to *paralinguistic modeling* we came up with two modeling approaches making use of static feature vector modeling describing statistics of acoustic speech features, e.g. pitch, intensity, and spectral characteristics. Ranking our features, we see that loudness and MFCCs seem most promising for the task of anger detection. For the SC-BAC database pitch features are likewise important. This might be an indicator for language differences between English and German when speakers express anger. Due to the lack of adequate data, an analysis comparing pitch in anger and nonanger between German and English has to be postponed to future work.

The first paralinguistic modeling P_{std} relies on approaches common in related work, while the second *extended* P_{ext} modeling describes the acoustic and prosodic properties of the speech signal in more detail by better capturing the dynamics in the signal, e.g. by applying a linear regression analysis. We could show that considerable performance gains are possible by applying this procedure. Particularly the MOB database, entailing longer utterances, benefits from this detailed approach. We should not fail to mention that the extended model demands an increased computational complexity due to the higher number of features; and although feature selection may reduce the number of features while maintaining the performance, many calculations such as DCT are required nonetheless. One option that might help to reduce the computational complexity while keeping the performance on a similar level is the fusion of the basic paralinguistic model with interaction- and context-information, which allows similar performance scores as the extended paralinguistic model.

Linguistic Modeling

In terms of linguistic modeling we applied probabilistic and entropy-based models for words and phrases, e.g. BOW, TF-IDF, SRI, and EmoSal. SRI and EmoSal clearly outperform the vector space models. Modeling phrases improves the scores slightly. Moreover, modeling with (erroneous) automatic ASR transcripts instead of error-free human transcripts still shows comparable emotion recognition scores as long as the WA does not drop below 40 %. We could show that our expectations on linguistic anger detection for IVR corpora could not be fulfilled as fusing paralinguistic and linguistic models only slightly improves. This may be primarily attributed to the comparably strong performance of the paralinguistic model, but also to the short length of the samples. As the short nature of the utterances is a typical characteristic for state-of-the-art IVR applications, we propagate that emotion recognition in such command-based IVR systems may abandon linguistic models as their contribution appears limited. It should be noted that this might not hold for data sets with considerable turn length and higher perplexities as shown by Schuller (2006).

Interaction- and Context-Related Modeling

We demonstrated clear dependencies between the interaction and the invoked emotions. Anger is provoked through malfunctioning ASR whereas by contrast a well-working interaction implies nonanger. This dependency could be shown by a correlation analysis plus an IGR ranking scheme and holds true for both analyzed data sets. System-dependent differences could be observed in the strength of the correlation. LG-FIELD shows stronger ties between interaction parameters and anger as SC-BAC. The stronger correlations of the interaction parameters with the emotional state in the LG-FIELD data account for the good performance of the interaction- and context-related model when applied on LG-FIELD. Furthermore, LG-FIELD users tend to get more annoyed with increasing dialog duration, whereas SC-BAC users are uninfluenced by the dialog progress. Notwithstanding this, the models generalize quite well.

History-Related Modeling

As statistics shows, emotions never occur in isolation but form upon a pattern. With the presented HMM-ensemble it could be shown that we may use this information to estimate the likelihood of observing an emotion based on previous emotional states. Similarly as in the interaction- and context-related approach we transcend utterance boundaries. The performance may vary greatly depending on the distribution of the emotions. For the MOB data set the implication that anger relates to further anger is stronger as in the SC-BAC data set, which is beneficial for predicting anger in MOB as a higher F-measure for the ANGER class could be observed. The current limitation of this approach is the employment of hand annotated emotion labels, which may be compensated by using a prior paralinguistic emotion recognizer. Considerable performance losses could not be observed when following this procedure, cf. Schmitt and Polzehl (2010). The detailed effect of erroneous automatic labels and additional emotion classes needs to be studied in future work.

Corpus Differences

Signal Quality While the AIBO database has been recorded in wide band quality under controlled conditions, the three IVR databases are of narrow-band quality including noise. However, signal quality does not seem to be the predominant factor in our anger recognition experiments since the results for the assumed higher quality AIBO database do not yield better classification scores. Furthermore, the IVR speech quality can also be subdivided into sub-classes, since callers may have dialed in via different transmission channels using different encoding paradigms. While the SC-BAC database mostly comprises calls that were routed through land line connections the MOB and LG-FIELD databases account for a greater share of mobile telephony transmission channels. Because fixed line connections usually transmit less compressed speech it could be assumed that there is more information retained in it. Eventually, the impact of the difference in the total amount of information between wide-band and narrow-band speech as well as the differences caused by speech transmission remain to be addressed in further experiments in the future.

Speech Duration Another factor in the database design is the average turn length, which also correlates to the number of words in a turn. On the one hand, a longer turn offers more data for both linguistic and acoustic analysis. On the other hand, emotional expression can be very short. Longer passages may include diverse emotional user state changes. In terms of modeling, statistics are presumably more reliable on the basis of longer speech utterances, as the acoustic systems are based on statistics derived from the entire turn. Also Vlasenko et al. (2008) concludes, that larger units seem to be beneficial for emotion recognition. Also, linguistic phrase modeling can be expected to improve. However, the differences in speech duration in the selected databases do not show significant influence. The MOB, AIBO, and LG-FIELD-databases are comparable in terms of average words per turn, though the standard deviation is higher for the IVR databases. The SC-BAC is of lower average. In the present work, differences in overall performance do not appear to be correlated to this condition. As speech duration is mostly limited by database-dependent turn lengths or phrase lengths, most common units of analysis are naturally turns or words. In contrast to this, modeling on sub-word level, e.g. phonemes, has been applied in Vlasenko and Wendemuth (2009) and on phoneme-classes in Bitouk at al. (2010). Eventually, the present diversity in the literature suggests that the optimal analysis unit strongly depends on the data. Segmentation and labeling issues need to be addressed in future work in this respect.

Class Labels and Agreement The inter-labeler agreement is similar for the MOB and AIBO databases with $\kappa = 0.53$ for the German IVR and $\kappa = 0.56$ for the German AIBO corpus. The slightly higher $\kappa = 0.63$ for the English SC-BAC corpus could indicate a clearer separation of the expressive patterns and could thus explain the good performance on the SC-BAC IVR. The low performance of the AIBO corpus can perhaps be attributed to the diversity of the anger class labels which comprise *touchy*, *angry*, *reprimanding*, and *emphatic*. Here, acoustic profiles may be overlapping, interfering, or mutually blurring. Experiments using the original class definitions may well lead to more precise results. Also differences in labeler training could be influential. Steidl (2000) concludes, the mere task of choosing exactly one category forces the labelers to come to hard and clear-cut decisions, independently of the intensity of the emotional state. The very point of cutting varies intra- and inter-personally. Still, the κ values achieved here are typical for emotion databases, which generally amount to "fair" agreement between labelers only. This points to a general difficulty in determining the ground truth for an abstract concept, such as "emotion", when only acoustic and linguistic evidence is available. This poses a challenge for all work on real-life data on this task, be it performed by humans, or machines.

Further Considerations

Once we consider deploying emotion recognition in SDS, we have to define the priority of how reliable we want the detection of certain emotions to be in relation to others. This priority depends on the actions taken by the DM as response to an emotional state since it has a direct repercussion on the dialog. It is worth reiterating at this point that general performance scores for assessing emotion recognizers such as the accuracy (which is used in large parts of related studies) are thus misleading. When classifying emotions, class-wise scores are far more important evaluation, since some emotional states, such as anger, may demand a higher recognition rate as others. It should be noted that in the binary anger recognition scenario particularly the anger class is less reliably detected as the nonanger class. This holds true for virtually all data sets when applying the paralinguistic models, the vector-based linguistic models, the interaction- and context-related, and the history-related models. Solely EmoSal and SRI show better F-measures for the anger class, which in any case had only minor performance gain on the fused system. Our response to this phenomenon has been the introduction of a cost-sensitive classifier. By altering the costs we have control over the precision of predicting ANGER, which in return is paid with a lower recall. Particularly for anger detection, greater efforts are required to increase the F-measure. Leaving out blurring patterns, such as SLIGHTANGER, when training classifiers might bring the solution.

Chapter 5
Novel Approaches to Pattern-Based Interaction Quality Modeling

In the previous chapters it has been shown that users show emotions towards SDS. However, it has also become obvious that only a minority of the users reacts emotionally while interacting whereas the vast majority often remains emotionless. This user behavior seems natural as users do not expect systems to be able to recognize emotions. While the presence of emotions may allow inferences to be drawn on user satisfaction, the absence of emotions consequently may not. A user may be dissatisfied without showing anger and satisfied without showing happiness, i.e., an interaction may be endangered without the user showing emotions. As a result, other (complementary) approaches to emotion recognition are required for online monitoring SDS in order to detect critical dialog situations.

In the following, we assume that the "quality" of an interaction may be estimated by exploiting the dependency of input variables representing the interaction, and target variables, representing the quality at arbitrary points in an interaction. Therefore, we describe a pattern-based statistical modeling approach with the aim of endowing SDS with the capability to notice if an interaction is going wrong. The proposed approach resembles the PARADISE paradigm, but differs in that it is suited to estimate online the quality at arbitrary points in the interaction.

In Sect. 5.1 we discuss the term *user satisfaction* (US) and motivate the expression *Interaction Quality* (IQ). We bring up for discussion whether it is more target-aimed to capture the users' opinions for modeling "quality" or expert annotators that pretend to be users. In the following, both approaches are pursued with the intention to provide a comparison of both target variables. Afterwards, we present the annotation scheme for IQ in Sect. 5.2 and define a set of guidelines that may be used by expert annotators when judging over the quality of an interaction. 200 dialogs from a real-life SDS with users interacting in the field *"Let's Go Field Corpus"* (LG-FIELD) are annotated by human experts. The optimum merging strategy of all expert annotations is further discussed. Moreover, the results of annotating the LG-FIELD data set with IQ labels are shown and analyzed. For comparing the congruency and correlation of expert annotations with subjective user impressions we conducted a user study, which is presented in Sect. 5.3. In this study, 46 users have been asked to interact

A. Schmitt and W. Minker, *Towards Adaptive Spoken Dialog Systems*,
DOI: 10.1007/978-1-4614-4593-7_5,

with the same SDS under laboratory conditions and to provide US scores during the interaction, resulting in the "Let's Go Lab Corpus" (LG-LAB). We can show that an objective measurement of IQ highly correlates with US. In Sect. 5.4, input parameter sets are defined that are used to predict IQ and US. They serve as input variables for the statistical models. Endowed with the input variables and the target variables IQ and US, we define a statistical model in Sect. 5.5 that allows an estimation of the quality.

In Sect. 5.6, the approach is evaluated using SVMs. The contribution of the input variables from the different system modules on this task are assessed and discussed and relevant features are identified. Using feature selection, we also show that the performance of the model may be further increased. That followed, the generalizability of the model is examined using different corpora for training and testing. Afterwards we turn our attention to the linear dependency between the input variables and IQ/US. First of all, a linear regression model is created and analyzed. Furthermore, correlations between the input variables and the target variables are presented. It will be further scrutinized, which system events directly caused the users to raise or lower their satisfaction rating. We conclude this chapter by summarizing and discussing the results in Sect. 5.7.

5.1 Interaction Quality Versus User Satisfaction

In the literature no rigorous definition of the term "user satisfaction" yet exists. According to Doll and Torkzadeh (1991) "user satisfaction" is the opinion of users about a specific computer application, which they use. Other terms for "user satisfaction" are common, such as "user information satisfaction", which is defined as "the extent to which users believe the information system available to them meets their information requirements" (Ives et al. 1983). Larcker and Lessing (1980) speak about "perceived usefulness of information" when addressing user satisfaction. User satisfaction and usability are closely interwoven. ISO (1998) subsumes under the definition "usability" a compound of efficiency, effectiveness and satisfaction. Yet satisfaction is often seen as a by-product of great usability in HCI literature (Lindgaard and Dudek 2003).

While the definition of "user satisfaction" implies a completely subjective metric, the term "quality" includes both, subjective and objective characteristics: The different aspects of quality criteria related to HCI are summarized under the terms *Quality of Service* (QoS) and *Quality of Experience* (QoE). QoS is defined as "the collective effect of service performance, which determines the degree of satisfaction of the user" (ITU 1994) and depicts a more objective view on quality. It can be determined by the system developer. In contrast to this stands QoE, which is defined as "the overall acceptability of an application or service, as perceived subjectively by the end-user." (ITU 2007). The definition implies that QoE requires a subjective judgment process by the user and can thus only be measured with user surveys (Moeller et al. 2009).

However, measuring QoE, i.e., subjective user judgments about the interaction, turns out to be difficult, basically for two reasons:

- *Strong variations*: It can be assumed that each user has a different understanding of a functioning interaction. Some might perceive a dialog behavior, e.g., a perpetual posing of the same question, as very disturbing whereas others would not even take notice of it. This would imply that tracking user satisfaction is user-specific and would require models trained on data from a specific user, cf. (Engelbrecht et al. 2009).
- *Trackability problem*: Tracking real user satisfaction online, i.e., throughout a spoken interaction with the aim to create user-specific models could only take place under laboratory conditions. The users would then pursue a fake task in an artificial environment. To which extent this scenario differs from a real-life scenario with disturbing environmental factors can hardly be examined. It should be further noted that a user would have to synchronously interact with a system plus adjusting his satisfaction. However, this depicts a high cognitive load for the test subject and in the worst case could falsify the user's judgment about the interaction.

The previous related work shows the challenge that persists in modeling user satisfaction on the exchange level: an ideal model would mirror the actual satisfaction of each individual user interacting with a system. This would require user-specific models, which basically cannot be realized for SDS serving a large number of users. The second best option is the implementation of general user satisfaction models that reflect the mean satisfaction of real users. Nevertheless, the trackability problem remains.

The less cost-intensive and in our view more objective metric is to employ expert raters, similarly as a third person observing the interaction between railroad clerk and customer (cf. Sect. 1.2). This allows the use of real-life data, which implies that the models are trained on data from scenarios they are intended to be deployed for. Expert annotations further exclude a user bias, e.g., general attitudes towards an SDS are faded out. The resulting model would then mirror a more objective view on an interaction and would thus yield a higher predictive performance.

We define "Interaction Quality" as an objective quality metric that quantifies the quality of the interaction between a system and the user up to a certain point in an interaction on a score from 1 to 5. The architecture of our IQ framework is depicted in Fig. 5.1. The metric relies on opinions from several expert raters that provide quality scores for the observed interactions. At points where a rater feels a user would be dissatisfied, due to bad system performance, the type of response received by the system, or simply the fact that the user seems acoustically annoyed, the raters are encouraged to assign low scores. Objectivity, and by that a universality of the model, is reached through a combination of the votes from the single raters plus a rule set, which assures that results are reproducible elsewhere. Interaction quality can be predicted by a set of input parameters and the emotional state of the user, which may be obtained by the emotion recognition approaches presented in Chap. 4.

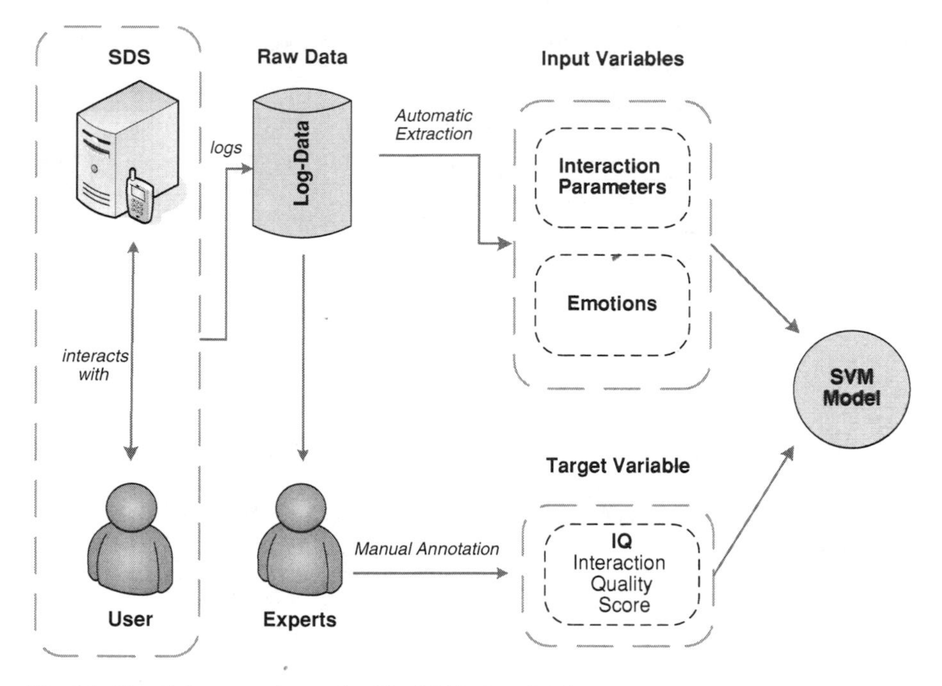

Fig. 5.1 The IQ framework, see also Fig. 2.11 in Sect. 2.3.1

5.2 Expert Annotation of Interaction Quality

Our main interest lies in the prediction of IQ in field scenarios. In the following, real-life data will be endowed with IQ labels, resulting in a data corpus that we will refer to in the following work as LG-FIELD. Three expert raters, advanced students of computer science and engineering familiar with SDS technology, annotated respectively 200 dialogs comprising 4,885 system-user exchanges from the CMU Let's Go bus information system recorded in 2006. The raters were asked to annotate the quality of the interaction at each system-user exchange with a score from "5" (very good), "4" (good), "3" (fair), "2" (poor), "1" (very poor). We have chosen a number of three raters considering this as a tradeoff between costs and subjectivity. The three different subjective opinions about the Interaction Quality can later be merged to a more objective score.

Higashinaka et al. (2010a, b) who conducted similar studies in the context of human-human and human-machine text-chat systems report low correlation among the ratings from several raters when annotating user satisfaction (Spearman's ρ 0.04–0.32). This motivated us to develop a set of basic guidelines that should be used by the raters, cf. Table 5.1.

The guidelines have been designed in such a way that the raters still have a sufficient level of freedom when choosing the labels. However, they should prevent them from too strong variations among the neighboring system-user exchanges due to

Table 5.1 Rater guidelines for annotating Interaction Quality (adapted to the CMU Let's Go domain)

Rule	Description
1	The rater should try to mirror the users point of view on the interaction as objectively as possible
2	An exchange consists of the system prompt and the user response. Due to system design, the latter is not always present
3	The IQ score is defined on a five-point scale with "1 = very poor", "2 = poor", "3 = fair", "4 = good" and "5 = very good"
4	The Interaction Quality is to be rated for each exchange in the dialog. The history of the dialog should be kept in mind when assigning the score. For example, a dialog that has proceeded fairly poor for a long time, should require some time to recover
5	A dialog always starts with an Interaction Quality score of "5"
6	The first user input should also be rated with 5, since until this moment, no rateable interaction has taken place
7	A request for help does not invariably cause a lower Interaction Quality, but can result in it
8	In general, the score from one exchange to the following exchange is increased or decreased by one point at the most
9	Exceptions, where the score can be decreased by two points are e.g. hot anger or sudden frustration. The rater's perception is decisive here
10	Also, if the dialog obviously collapses due to system or user behavior, the score can be set to "1" immediately. An example therefore is a reasonable frustrated sudden hang-up
11	Anger does not need to influence the score, but can. The rater should try to figure out whether anger was caused by the dialog behavior or not
12	In the case a user realizes that he should adapt his dialog strategy to obtain the desired result or information and succeeded that way, the Interaction Quality score can be raised up to two points per turn. In other words, the user realizes that he caused the poor Interaction Quality by himself
13	If the system does not reply with a bus schedule to a specific user query and prompts that the request is out of scope, this can nevertheless be considered as "task completed". Therefore this does not need to affect the Interaction Quality
14	If a dialog consists of several independent queries, each query's quality is to be rated independently. The former dialog history should not be considered when a new query begins. However, the score provided for the first exchange should be equal to the last label of the previous query
15	If a constantly low-quality dialog finishes with a reasonable result, the Interaction Quality can be increased

individual conditions. By all means, the rule set should not provide a recommendation to the rater that instructs them when to assign which score, since this would reduce a statistical model to absurdity. According to the guidelines every dialog is initially rated with a score of "5" since every interaction at the beginning can be considered as good until the opposite eventuates. Our model assumes that users are initially interacting with an SDS without bias, i.e., the basic attitude toward a dialog system is assumed to be neutral. Other assumptions would not be statistically predictable.

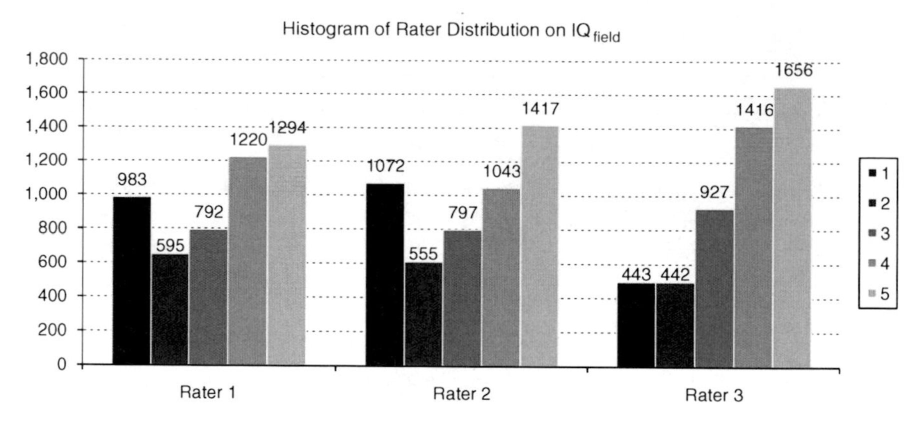

Fig. 5.2 Rating distribution for Interaction Quality within the Let's Go Corpus for each rater

5.2.1 Annotation Example

An example dialog is depicted in Table 5.2 along with the ratings from three expert annotators. As can be seen, the raters initially consider the IQ as good until exchange number seven, where it is obvious that the system has misunderstood the user. Rater 3 deviates from the IQ scores provided by the other two raters at this point and lowers his IQ score already at turn number five. This might be due to the fact that the rater feels that the system delivers a too detailed prompt when explaining the possible bus routes at turn number four. In general, the raters 2 and 3 assign the lower scores compared to rater 3 throughout the entire interaction. A certain subjectivity of IQ is already visible in the chart and it can be further noted that the median determined from all three rater's opinions smoothes the subjective scores, leading to a more objective view on IQ. The motivation for the median as final label is discussed later in this chapter. The lower chart of Table 5.2 is a look ahead of the IQ model introduced later in this chapter (cf. Sect. 5.5). It contains the hypothesis of the statistical model in relation to the reference median label. It is depicted here only for comparison reasons.

5.2.2 Rating Statistics and Determination of the Final IQ Score

The distribution of the labels provided by the single raters is shown in Fig. 5.2.

As expected, the distribution is skewed toward label "5" since every dialog initially is assumed to have a good IQ. The varying distributions in the histogram and the inter-rater agreement shows that Interaction Quality is still a subjective metric, although guidelines seem to synchronize the labels to a greater extent than when no guidelines are applied. The overall mean agreement can be reported with Cohen's $\kappa = 0.31$

Table 5.2 Example dialog (ID: 2061122025) from the CMU Let's Go System (2006 corpus) with low Interaction Quality

1	Welcome to the CMU Let's Go bus information system
2	To get help at any time, just say Help or press zero
3	What can I do for you? $\langle DTMF - Zero \rangle$
4	I am an automated spoken dialog system that can give you schedule information for bus routes in Pittsburgh's East End. You can ask me about the following buses: 28X, 54C, 56U, 59U, 61A, 61B, 61C, 61D, 61F, 64A, 69A, and 501
5	What bus schedule information are you looking for?
6	For example, you can say, when is the next 28X from DOWNTOWN to THE AIRPORT? or I'd like to go from MCKEESPORT to HOMESTEAD tomorrow at 10 a.m. *28X*
7	The 56C. Is this correct? *No, 28X*
8	What can I do for you? *I want to know whether /uh/ tomorrow the bus is available or not*
9	Leaving at 11 p.m. Did I get that right? *No, I just want to know is there any bus service for 28X tomorrow*
10	Please use shorter answers because I have trouble under standing long sentences
11	Leaving at 11 p.m. Is this correct? *On twenty third november*
12	Leaving at 11 p.m. Is this correct? *No, I want to know the schedule on* twenty third november
13	Leaving at 2 a.m. Did I get that right? *< noise >*
18	Leaving from DYNAMO. Is this correct? *No*
19	Leaving from DYNAMO. Is this correct? *Leaving to airport*
20	Leaving from AIRPORT . Did I get that right? *yes*
21	Right
22	Where are you going? *To airport*
23	Going to AIRPORT. Did I get that right? *yeah, yep*
24	Okay
25	When would you like to travel? *yes*
26	When would you like to travel? *Twenty third November.*
27	When would you like to travel? *Tomorrow*
28	At what time do you wanna travel? *Seven a.m.*
29	Leaving at 7 a.m. Is this correct? *yep*
30	Alright
31	Just a minute. I'll look that up
32	AIRPORT and AIRPORT are both the same stop. Please provide a different start or end point.
33	To ask about a different trip, you can say, start a new query.If you are finished, you can say goodbye *Yes*
34	You can say start a new query, or goodbye *No*
35	To ask about a different trip, you can say, start a new query If you are finished, you can say

(continued)

Table 5.2 (continued)

14	If you want to leave at 2 a.m. say yes or press one, otherwise say no or press three *no < noise >*		goodbye *< noise >*
15	Going to NORTH AND CEDAR. Did I get that right? *28x going to airport at seven a.m.*	36	You can say start a new query, or goodbye *< noise >* You can say start a new query, or goodbye. *< hangup >*
16	Please use shorter answers because I have trouble understanding long sentences.		
17	Going to NORTH AND CEDAR. Is this correct? *No*		

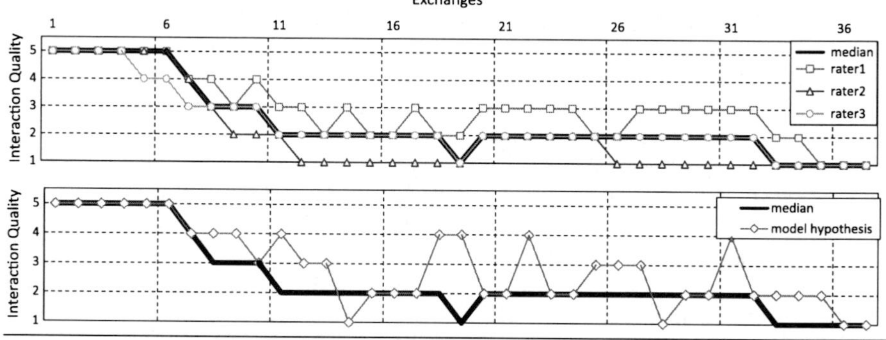

The user utterances are printed in italic

Upper chart Turn-wise IQ annotation from 3 raters. The final label is the median of all three opinions. *Lower chart* median reference versus hypothesis of the model trained with AUTO feature set

(Carletta 1996), which can be considered as "fair agreement" according to Landis and Koch (1977). The correlation among the raters can be reported with Spearman's $\rho = 0.72$ (Spearman 1904), which depicts a by 0.40 points higher correlation as reported by Higashinaka et al. (2010a).

Since the aim is to model a general opinion on Interaction Quality, i.e., it should mirror the IQ score other raters (and eventually users) agree with, the final label is determined empirically. A majority voting for the distinction of the final label has to be excluded since in 21 % of the exchanges all three raters opted for different scores and a majority vote would not be applicable here. Thus we consider the mean of all rater opinions as possible candidates for the final class label:

$$\text{rating}_{\text{mean}} = \left\lfloor \left(\frac{1}{R} \sum_{r=1}^{R} \text{IQ}_r \right) + 0.5 \right\rfloor, \tag{5.1}$$

Table 5.3 Agreement of single rater opinions to the merged label when determined by mean and median, measured in κ, ρ and accuracy

	Mean label	Median label
Cohen's κ		
Rater1	0.557	0.688
Rater2	0.554	0.679
Rater3	0.402	0.478
Mean	0.504	**0.608**[a]
Spearman's ρ		
Rater1	0.901	0.900
Rater2	0.911	0.907
Rater3	0.841	0.814
Mean	**0.884**	0.874
Accuracy		
Rater1	0.651	0.755
Rater2	0.647	0.749
Rater3	0.539	0.598
Mean	0.612	**0.701**[a]

[a] significantly higher ($p < 0.05$)

where IQ_r is the IQ score provided by rater r. $\lfloor y \rfloor$ denotes the biggest integer value smaller than y. Every value IQ_r contributes equally to the result that is finally rounded half up to an integer value. Furthermore we consider the median, which we define as:

$$\text{rating}_{\text{median}} = \text{select}(\text{sort}(IQ_r), \frac{R+1}{2}), \tag{5.2}$$

where sort is a function that orders the ratings IQ_r of all R raters ascendingly and select chooses the item with index i from the list. The compliance of the single user ratings with the final label (calculated on mean and median) is depicted in Table 5.3. Cohen's κ, Spearman's ρ and the accuracy, i.e., the percentage of agreement, are applied.

As can be seen, the agreement of the three raters with the median label is significantly[1] higher than with the mean label. Consequently, the median label depicts the broadest consensus and represents the most objective measurement of IQ. It therefore commends itself for creating the model.

5.3 A User Satisfaction Study Under Laboratory Conditions

A central advantage of IQ is that it can be easily reproduced to model the quality in other systems and system versions without the need for user tests. However, it has not

[1] Determined with a paired two-sample t-test.

been clarified to which extent IQ correlates with user satisfaction, i.e., to which degree IQ mirrors the user's perception of the interaction. In the following, we present a lab study that targeted on measuring user satisfaction during SDS interactions, which resulted in the LG-LAB. The study sheds light on the dependency of IQ and US and provides baselines for predicting US.

5.3.1 Lab Study Setup

For evaluating a basis for comparison with the previously annotated field data, the user study was conducted with the same system that depicts the source for the field data. The study was conducted with the Let's Go Bus Information system, which is publically available since the Spoken Dialog Challenge (Black and Eskenazi 2009). The system has been set up in a university lab. Satisfaction level tracking at any point in time during an interaction implied that the users had to focus on both, the interaction and the user satisfaction adjustment. An increased cognitive load can be assumed in comparison to an interaction without parallel satisfaction tracking. We thus decided against implementing the feedback mechanism on a regular desktop PC. We assume that this could have distracted their concentration on focusing on the spoken dialog. Instead, we realized the feedback on a stand-alone, programmable LCD display with a touch screen. By that, we aimed to direct the user's attention on the task. Furthermore, the user feedback mechanism was provided using a presenter device that contained two large buttons "up" and "down", which are usually intended for skipping between presentation slides. In the test scenario the buttons have been used to decrease and increase the user satisfaction score. Alternatively the users could employ the touch screen, which, however, turned out to be less intuitive. Most participants decided to use the presenter after an initial test run. We further decided against the use of a handset and employed a headset with integrated microphone to additionally limit the cognitive load of the participants, see Fig. 5.3a. Moreover, the participants were placed in an isolated corner of the lab, with the investigator sitting out of sight, but in the same room. The input and output devices are depicted in Fig. 5.3b.

For proper input tracking the survey devices have been monitored by a newly developed survey management system that is temporally synchronized with the Let's Go system. Each change in satisfaction triggered by the user was logged, see Fig. 5.4. The latest change during a system-user exchange was taken as satisfaction score.

Our test persons followed five scenarios, that have been selected randomly from five predefined ones. In four scenarios the task was solvable, whereas in one scenario the Let's Go system is unable to provide a schedule. A solvable scenario is depicted in Fig. 5.5, the remaining scenarios are listed in Appendix B.3.

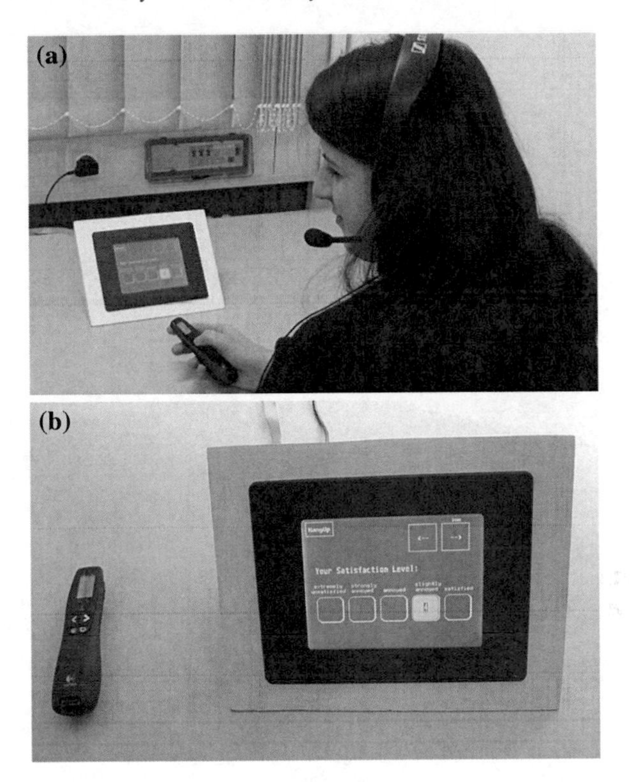

Fig. 5.3 Laboratory test setup for user satisfaction study. **a** Test person interacting with th Let's Go Bus Information System under laboratory conditions. **b** Input and output devices for adjusting user satisfaction. *Left* presenter input device. *Right* LCD display providing feedback

5.3.2 Participants

In total, 46 subjects (22 female, 24 male) participated in the study. The majority of the participants, aged between 19 and 55 years, with a mean of 28.7 years, have been recruited from the university campus, and some have been non-university members. Each test person was rewarded with a cinema voucher for participating. The vast majority have been students or postgraduates. Altogether 40 % indicated a technical background. Every subject uses computers in everyday life with a daily usage time between 15 and 600 min (262 min in average). 66 % did have prior experience with SDS, either as a customer or job-related, with 57 % of them declaring their experience as positive. Hence, the users can be considered as very familiar with technology and adept in using computers. All participants have learned English as the first foreign language while German was the mother tongue (86 %). Even though no English native speakers participated the average English skills can said to be high, since 81 % quoted it to be "fluent" or "business fluent" while only one participant declared "basic" knowledge. To facilitate the comprehensibility of the system prompts, we

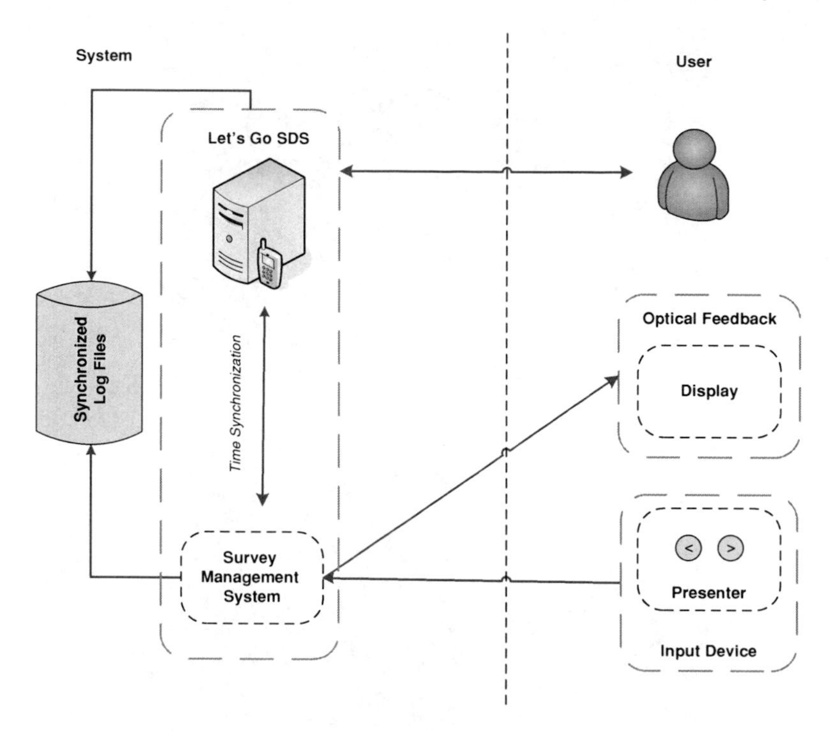

Fig. 5.4 Schematic setup of the user satisfaction study

replaced the standard TTS voice from the Let's Go system with a more sonorous TTS voice from Cepstral.[2] The output has further been slightly decelerated by 20 %. This step also gave the participants more time to consider their satisfaction ratings during the interaction.

Test Procedure

Each user filled out a questionnaire measuring the technical affinity and further user characteristics. Next they were instructed to conduct five phone calls according to the five predefined scenarios in random order. The users have been instructed to interact with the system without bias, i.e., prior experiences with other dialog systems or experiences with one of the previous dialog scenarios should be faded out. Furthermore, the users were told that a "help" functionality exists in the system, but that its use is optional. They were further asked to hang up if they would do so in reality.

[2] Cepstral William from www.cepstral.com.

Fig. 5.5 Test-case scenario

Task 1
You are on a business-trip in Pittsburgh, PA and you have just arrived at the airport. The hotel where you will stay is in downtown Pittsburgh. You decide to go there by bus, but you don't have any schedule information about the bus system there. You want to call the Let's Go Bus Information System to obtain this information. The first meeting will start soon, so you want to check in into your hotel as quickly as possible.

Please try to obtain schedule information for a bus route from the *Airport* to *Downtown Pittsburgh* for the *next bus* and the one *after* that.

The user satisfaction scheme has been chosen to be comparable with the Interaction Quality scores, i.e. the ratings range on a linear scale from "very satisfied" (5), "satisfied" (4), "somewhat dissatisfied" (3), "dissatisfied" (2), "very dissatisfied" (1).

Consequently, the user has intentionally no possibility to adjust a higher satisfaction than "very satisfied". Furthermore, very satisfied, i.e., a score of "5" is the initial satisfaction score that is assumed. While this may seem odd on the first sight, it allows for a comparison of the both IQ and US scales. Similarly to the IQ instructions, the users were asked to mirror their *overall* satisfaction with the *entire previous interaction*, i.e., with the current score the user should not rate the satisfaction with the last step. After each call was finished, the users were asked to fill out a short questionnaire on the touch screen. The overall satisfaction with the dialog has been tracked. Furthermore, we asked them to indicate when they felt that they did not completely mirror their satisfaction at all times for being able to sort out invalid dialogs.

Data Summary

The collected data comprised 230 calls. It was reduced to a set of 128 originating from 38 different users. Calls where the system crashed, which happened from time to time, have been sorted out. Furthermore, we excluded calls where the user stated of not having provided satisfaction scores in time. The statistics of the final corpus is depicted in Table 5.4.

Table 5.4 Statistics of LG-LAB

	Total	AVG	SD
Users	38	–	–
Calls	128	3.37	0.79
Pairs of system/user turns	2,988	23.34	10.50

AVG and SD denote the average number and the standard deviation of calls per user and the number of pairs of system/user turns per call

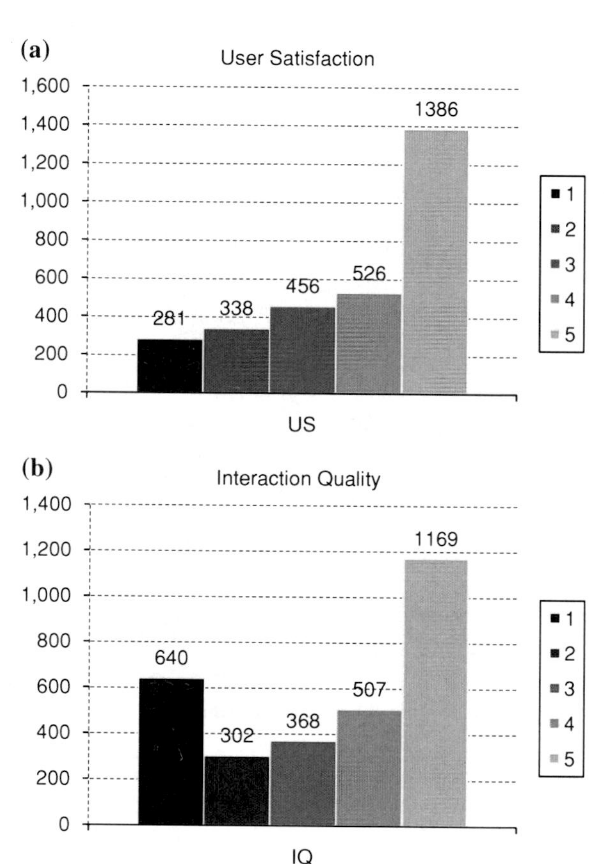

Fig. 5.6 Distribution of the US and IQ ratings in LG-LAB for all system-user exchanges

5.3.3 Comparison of User Satisfaction and Interaction Quality

The collected data has been manually annotated with IQ scores as previously described for LG-FIELD. Figure 5.6 depicts the distribution of IQ and US labels in LG-LAB.

It is interesting that the US score exhibits a higher amount of high user satisfaction ratings US_{lab}, i.e., a score of "5", than the IQ_{lab} score. Either, the users are obviously more satisfied than the expert raters assume, or some forgot to lower the score when being dissatisfied. The latter should, however, not occur since dialogs have been

Table 5.5 Agreement (κ) and correlation (ρ) in IQ ratings of the 3 raters in LG-FIELD and LG-LAB plus agreement and correlation of IQ_{lab} with US_{lab}

LG-FIELD					
	R1/R2	R1/R3	R2/R3	Mean	
κ	0.41	0.25	0.26	**0.31**	
ρ	0.79	0.68	0.70	**0.72**	
LG-LAB					
	R1/R2	R1/R3	R2/R3	Mean	IQ_{lab}/US_{lab}
κ	0.42	0.24	0.22	**0.29**	**0.31**
ρ	0.79	0.59	0.63	**0.67**	**0.66**

Objective expert ratings (IQ_{lab}) show similar correlations and agreement values with user satisfaction US_{lab} as the expert raters among each other (see mean values)

previously sorted out where the users stated to have forgotten the US adjustment. It can be further noted that the IQ_{lab} score displays a higher amount of very dissatisfied turns, i.e., a score of "1". The higher occurrence of "1" in IQ_{lab} instead of "2", "3" and "4" can be explained by the fact that many calls with low IQ_{lab} end up with a disastrous interaction that can be considered as non-reversible. The users in the study obviously had a more positive perception on such negative interactions. It is further remarkable that the mean US_{lab} amounts to 3.80, whereas the mean IQ_{lab} only amounts to 3.43, which depicts a highly statistically significant difference.[3] Moreover, it can be noted that the distribution of IQ_{lab} is equally shaped as the IQ_{field} annotations, which indicates a reproducible annotation scheme. When looking at the correlation and inter-rater agreement of the single raters in both corpora LG-FIELD and LG-LAB in Table 5.5 we can note that the values among the raters are comparable to those of raters and users. In other words: the degree of agreement and correlation for experts vs. experts is comparable to experts vs. users. The correlation of US_{lab} and IQ_{lab} is strong with $\rho = 0.664$ and is also highly statistically significant.[4] Further, the agreement of users and the raters amounts to $\kappa = 0.307$ and depicts a comparable value as the individual raters in the IQ_{field} annotation task. IQ raters and users seem to agree on a large proportion of the labels. If the users denoted "1" as satisfaction, the raters agreed in 67 % of the cases, if the users rated "5", the raters even agreed in 83 %.

5.4 Input Variables

In the previous section, the two corpora have been endowed with the target variables IQ_{field} and IQ_{lab}, as well as US_{lab}. For a prediction of the target variables, a set of input variables is required, representing relevant information about the interaction between the system and the user. For use on the dialog level, a comprehensive set of

[3] Determined with a paired two-sample t-test ($p < 0.01$).

[4] Determined with a paired two-sample t-test ($p < 0.01$).

such parameters has been defined in the ITU Supplement 24 to P Series "Parameters describing the interaction with spoken dialog systems" ITU (2005). It summarizes the interaction for *entire* dialogs and can be employed for *offline* classification. In Sect. 3.2.1, a set for quantifying and describing the interaction at a specific exchange suitable for *online* classification has been introduced. Based on that set, we define the input variables describing the interaction.

Automatic Features
Summarized as "automatic features", we used a set of interaction parameters that may be derived from interaction logs without manual intervention. They originate from the ASR, SLU and DM modules. The ASR parameters describe performance values and variables from the automatic speech recognition module. They summarize how reliable the ASR recognizes the spoken user utterance. It can be assumed that ASR performance has a significant impact on user satisfaction, cf. Möller (2005b). Furthermore, the ASR parameters track if the user interrupted or did not respond to a system question as well as the modality the user chose to interact with the system. The employed parameters are listed in detail in Table A.1.

In short, the group "ASR" used for predicting IQ and US consists of the following features:

ASR ASRCONFIDENCE, ASRRECOGNITIONSTATUS, BARGE-IN?, EXMO, GRAMMAR, MEANASRCONFIDENCE, MODALITY, TRIGGERED GRAMMAR, UTD, UNEXMO?, UTTERANCE, WPUT, #BARGE-INS, #ASR REJECTIONS, #ASRSUCCESS, #TIMEOUTPROMPTS, #TIMEOUTS_ASRREJ, #UNEXMO, %BARGE-INS, %TIMEOUTS_ASRREJ, %ASRSUCCESS, %TIMEOUTPROMPTS, %UNEXMO, {MEAN}ASRCONFIDENCE, {#}BARGE-INS, {#}ASRREJECTIONS, {#}ASRSUCCESS, {#}TIMEOUTPROMPTS, {#}TIMEOUTS_ASRREJ, {#}UNEXMO.

The second set of features is derived from the SLU module that extracts the semantic meaning from the automatically transcribed user utterance. Details are listed in Table A.2. The group "SLU" consists of the following features:

SLU SEMANTICPARSE, HELPREQUEST?, {#}HELPREQUESTS, #HELPREQUESTS, %HELPREQUESTS, OPERATORREQUEST?, {#}OPERATOR REQUESTS, #OPERATORREQUESTS, %OPERATORREQUESTS.

As central unit in a dialog system the dialog manager keeps track of the interaction and provides information such as the current dialog step the system is in, the time elapsed since the beginning of the dialog, the number of tries to fill the semantic slot etc. The DM-related parameters are listed in Table A.3.

The group "DM" consists of the following features:

DM ACTIVITYTRIGRAM, ACTIVITYTYPE, ACTIVITY, DD, LOOP NAME, PROMPT, REPROMPT?, ROLEINDEX, ROLENAME, WPST, #EXCHANGES, #REPROMPT, #SYSTEMQUESTIONS, #SYSTEMTURNS, #USERTURNS, %REPROMPT, {#}REPROMPT, {#}SYSTEM QUESTIONS.

It should be noted that textual parameters such as PROMPT, UTTERANCE and SEMANTICPARSE in this setting are employed "as is", i.e., no vector-space representation is introduced, such as bag-of-words, in order to keep the size of the feature space manageable.

Hand- and Semi-Annotated Features

To provide comparability to previous work (Higashinaka et al. 2010a), we further introduce a dialog act feature group that we create semi-automatically:

DACT SYSTEMDIALOGACT: one of 28 distinct dialog acts, such as *greeting, offer_help, ask_bus, confirm_departure, deliver_result, etc.* USERDIALOGACT: one of 22 distinct DAs, such as *confirm_departure, place_information, polite, reject_time, request_help*, etc.

To assess the impact of visible emotions we further introduce the negative emotional state of the user that is manually annotated by a human rater who chooses one of the four different labels for each single user turn, see also Sect. 3.2.2:

EMO EMOTIONALSTATE: emotional state of the caller in the current exchange. One of *garbage, non-angry, slightly angry, very angry.*

From all 4,832 user turns in LG-FIELD, 68.5% have been marked as non-angry, 14.3% slightly angry, 5.0% very angry and 12.2% contained garbage, i.e., non-speech events. In LG-LAB the following distribution could be observed: 76.23% non-angry, 16.5% slightly angry, 1.3% very angry and 5.9% garbage. In principle, the emotion feature can be determined automatically by using emotion recognition as described in Chap. 4.

The *User* feature set comprises user-specific information derived from the survey. They are only available in LG-LAB and are employed to determine their contribution on the classification task.

USER AGE: The age of the user. GENDER: The gender of the user. COMPUTERDAILYMIN: The user's average time spent with computers per day. SDSATTITUDE: User's answer to the question "I prefer talking to a human instead of a Spoken Dialog System." from 1: "strong agree" to 5: "strong disagree".

5.5 Modeling Interaction Quality and User Satisfaction

To estimate IQ (and US) at arbitrary points in a spoken dialog we apply a statistical model.

Let exchange

$$\varepsilon = \{\pi_1, \ldots, \pi_n\} \tag{5.3}$$

be a dialog exchange formalized and quantified by interaction parameters, dialog acts, the emotional state and user information denoted as

$$\pi \in \{\text{ASR, SLU, DM, DACT, EMO, USER}\}, \tag{5.4}$$

where ASR, SLU and DM represent the interaction parameters including the window- and dialog-level parameters,[5] DACT the user and system dialog act, EMO the emotional state of the user and finally static user properties, denoted as USER. Further let

$$\theta = \{\{\varepsilon_1, \rho\}, \ldots, \{\varepsilon_j, \rho\}\} \tag{5.5}$$

be a training set of size j consisting of exchanges ϵ_i with one of the ratings $\rho \in \{1, \ldots, 5\}$. Then a statistical classifier can be trained with the training set θ resulting in a model μ. To determine the IQ score at a specific exchange j, we determine the feature vector ε_j and estimate

$$\widehat{\text{IQ}} = \arg\max_{\text{IQ}} P(\text{IQ}|\varepsilon_j, \mu). \tag{5.6}$$

Estimating $\widehat{\text{US}}$ is defined analogously. $\widehat{\text{IQ}}$ and $\widehat{\text{US}}$ may be determined with discriminative classification, i.e., the score predicted by the classifier is not considered as a continuous one but as distinct class $\in \{1, 2, 3, 4, 5\}$.

5.6 Evaluation

The model for estimating $\widehat{\text{IQ}}$ and $\widehat{\text{US}}$ is constructed with an SVM with linear kernel that uses the fast sequential minimal optimization (SMO) algorithm (Platt 1999). Input variables are features from the described groups, i.e., $\pi \in \{\text{DACT, ASR, SLU, DM, EMO, USER}\}$. The target variables considered here are $y \in \{\text{IQ}_{\text{field}}, \text{IQ}_{\text{lab}}, \text{US}_{\text{lab}}\}$.

5.6.1 *Performance Metrics*

The skew distribution of the five classes 1–5 requires the employment of an evaluation metric that weights the prediction of all classes equally. Hence, a performance metric, such as *accuracy*, would not be a reliable measurement. We selected the *unweighted average recall* (UAR) to assess the model's performance, see Sect. 2.1.2. Although it does not consider the severity of the error, i.e., predicting "1" for an IQ of "5" is considered as fatal as predicting "4", it has been proven to be superior to other evaluation metrics, see (Higashinaka et al. 2010a), where the UAR is called *Match Rate per Rating* (MR/R). It is defined as follows:

$$\text{MR/R} = \text{UAR} = (\mathbf{R}, \mathbf{H}) = \frac{1}{K} \sum_{r=1}^{K} \frac{\sum_{i \in \{i|R_i = r\}} \text{match}(R_i, H_i)}{\sum_{i \in \{i|R_i = r\}} 1}, \tag{5.7}$$

[5] i.e., parameters with prefixes {#}, #, $\{Mean\}$, $Mean$

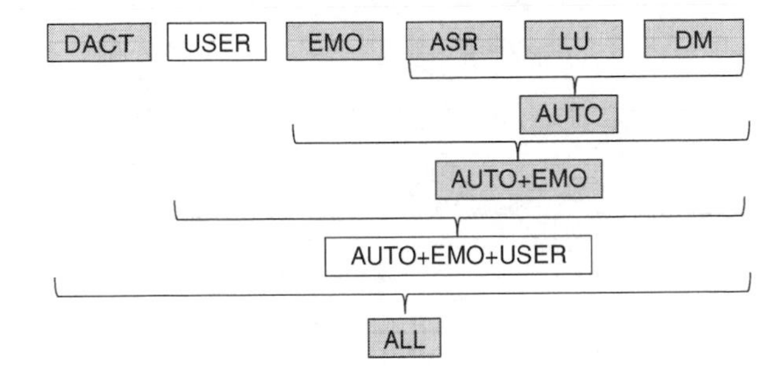

Fig. 5.7 Feature set composition of analyzed feature groups for predicting IQ_{field}, IQ_{lab}, US_{lab}

where K is the number of classes, here "5", and "match" is either "1" or "0" depending on whether the classifier's hypothesis H_i for the class r matches the reference label R_i. In the course of this chapter we will stick to the expression MR/R for clearness reasons. We further list Cohen's κ and Spearman's ρ to make our work comparable to other (future) studies but will use MR/R as central evaluation criterion and for optimization.

5.6.2 Feature Set Composition

In order to assess the performance contribution of the single feature groups, all available parameters have been grouped into nine different parameter sets, which are DACT, ASR, SLU and DM, see Fig. 5.7.

Furthermore, we subsumed the groups ASR, SLU and DM as AUTO set since all features of these groups can be automatically derived from logs without manual intervention. The AUTO + EMO group contains all AUTO features plus the emotion label. This will allow us to analyze the impact of emotions on classifying IQ_{field}, IQ_{lab}, US_{lab}. Moreover the AUTO + EMO + USER set additionally contains information from the USER group. It will show to which extent information about the user can improve classification. It should be noted that this set is only available for LG-LAB. Finally, all available features in LG-FIELD are summarized as ALL_{field}, containing the AUTO + EMO features plus the DACT features. The ALL_{lab} contain all parameters from LG-LAB.

5.6.3 Feature Selection

All available data has been split into two disjoint subsets consisting of 60 % of the dialogs for training and testing via stratified 10-fold cross-validation and the remaining 40 % of the dialogs for optimization. The dialogs have been selected randomly. This procedure has been applied for both corpora, LG-FIELD and LG-LAB with the target variables IQ_{field}, US_{lab}, IQ_{lab}.

Table 5.6 Top 10 features for IQ and US on optimization set according to IGR

	IQ_{field}		IQ_{lab}		US_{lab}	
	Feature	IGR	Feature	IGR	Feature	IGR
1	#ASRREJECT	1.0	#ASRSUCCESS	1.0	TURNNUMBER	1.0
2	#ASRREJECT_TIMEOUT	0.967	#USERTURNS	0.95	#SYSTEMTURNS	1.0
3	#ASRSUCCESS	0.834	#REPROMPTS	0.94	#ASRREJECT	0.96
4	#REPROMPTS	0.805	DD	0.94	#ASRSUCCESS	0.90
5	%REPROMPTS	0.800	TURNNUMBER	0.93	{MEAN}ASRCONFIDENCE	0.73
6	#TIME- OUTPROMPTS	0.757	#SYSTEMTURNS	0.93	ASRCONFIDENCE	0.68
7	#SYSTEMQUESTIONS	0.757	ASRCONFIDENCE	0.73	UTD	0.68
8	ROLEINDEX	0.699	UTD	0.73	#TIMEOUTPROMPTS	0.68
9	DD	0.566	%REPROMPTS	0.70	#TIMEOUTS_ASRREJ	0.68
10	#BARGE- INS	0.566	%ASRREJECT	0.68	%REPROMPTS	0.68

To determine the relevant features and to avoid the Hughes Effect (Hughes 1968), we perform feature selection on the optimization set and reduce the size of the feature space. The features are ordered according to an Information Gain Ratio (IGR) ranking (Witten and Frank 2005). The 10 most relevant features according to IGR for predicting IQ_{field}, IQ_{lab}, US_{lab} are depicted in Table 5.6.

As can be seen IQ and US are obviously heavily influenced by the performance of the ASR. It can thus be assumed that both the raters and the users tie their scoring to the ASR performance. Relevant variables are #ASRREJECT, #ASRREJECT_TIMEOUT and #ASRSUCCESS. Related to the capability of the ASR to recognize the user utterance, is the number of reprompts, i.e., the number of times the questions had to be prompted again since the ASR did not deliver a valid parse. The high rank of the variables #REPROMPTS and %REPROMPTS show that this behavior highly influences the IQ score, whereas the REPROMPT parameters for US seem to play a minor role. Interaction parameters that indicate the progress of the dialog are further relevant, such as the #SYSTEMQUESTIONS, #USERTURNS the TURNNUMBER and the dialog duration DD. This can be easily explained since the probability of observing low scores in the beginning of a dialog is comparatively lower than encountering low scores at later parts in a dialog. It is further obvious that parameters that contain running tallies, such as the previously discussed #ASRREJECT, #ASRREJECT_TIMEOUT and #ASRSUCCESS parameters are highly ranked as with increasing dialog progress the probability of observing higher values likewise increases.

For LG-LAB, the UTD further plays a dominant role. It seems that longer user prompts imply lower satisfaction. UTD weakly negatively correlates with IQ_{lab} and US_{lab} respectively with $\rho = -0.056$ and $\rho = 0.114$.

All features in the ranking belong to the group AUTO, i.e., they can be determined automatically during runtime. The hand-annotated (EMOTIONS) and semi-automatic features (SYSTEMDIALOGACT, USERDIALOGACT) obviously do not contribute as much as expected, since they do not occur among the top ten features. Furthermore, nearly all features are related to the overall interaction, i.e., features related to

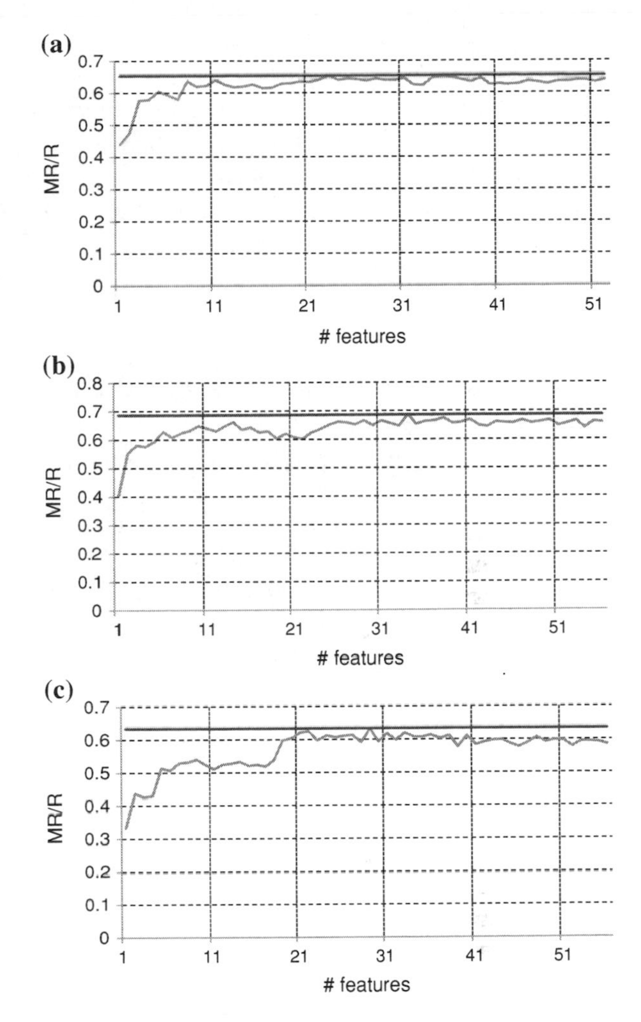

Fig. 5.8 Performance of the SVM when iteratively increasing the size of the feature vector with the k topmost features according to IGR on the optimization set and the ALL feature set. **a** Let's Go Field Corpus: IQ_{field}. **b** Let's Go Field Corpus: IQ_{Lab}. **c** Let's Go Lab Corpus: US_{Lab}

the current exchange, such as UTTERANCE, ASRSUCCESS?, etc. do not even occur. ASRCONFIDENCE and UTD and the window parameter {MEAN}ASRCONFIDENCE are the only exception.

To determine the global maximum of the classifier, i.e., the best performing feature set, we incrementally select the k topmost features from the list and perform 10-fold cross validation on the optimization set. Plots of the iterative feature selection on the ALL feature set are depicted in Fig. 5.8.

Several observations can be made: in all recognition tasks, a global performance maximum can be reached using only a very limited number of features from the entire set. It should be noted that all depicted plots are based on the optimization sets. For the IQ_{field} variable the best performing feature set consists of the top 23 features reaching an absolute performance of 65.3 % MR/R. However, a similar performance can already be gained with the 7 top-most features. All other features obviously neither significantly decrease nor increase the performance and can be considered irrelevant for predicting the IQ_{field} score on the optimization set. A similar behavior can be noticed when classifying the IQ_{lab} variable, which can be best predicted with the top 34 features, reaching 68.6 % MR/R. US_{lab} requires the top 29 features to reach a maximum performance of 63.4 %. It seems that the parameters beyond the *k* best performing features in all target variables neither harm nor contribute to the performance: the performance line beyond the maximum shows some jitter but does not considerably decrease when additional features are used.

5.6.4 Assessing the Model Performance

In a first step, all single feature groups have been evaluated including *all features* of a group using the training/test set. The evaluation can be considered as virtually user-independent since 10-fold cross validation using linear sampling has been applied.In a second step, the potential of feature selection on this task was examined. Therefore, the IGR procedure previously described, namely using the best-performing features according to IGR ranking and optimization process, has been repeated for all features of each single feature group. It should be noted that in this case the model has been also evaluated using the training/test set but the feature set has been reduced according to the results of the optimization set.

Results are depicted in the Table 5.7. The impact of feature selection on the model with its single feature groups can be seen in the lower part of the table.

IQ_{field} Performance

We will first direct our attention to IQ_{field} without feature selection. As can be seen, the model reaches a similar performance as Higashinaka et al. (2010a) with MR/R = 26.0 %, when trained with dialog act features alone. The slightly higher performance of our model can potentially be explained by the lower number of classes (five vs. seven), a different definition of the dialog act set, the employment of SVMs instead of HMMs or the difference in the target variable (IQ vs. closeness/smoothness/willingness). It can be observed that the utilization of other features considerably outperforms dialog act features. Particularly the group of the ASR features alone reaches a performance of MR/R = 60.5 %. The employment of all features delivers MR/R = 58.4 %, which is 1.8 % below the ASR features. We assume that here the too large number of (irrelevant) parameters causes this lower performance. Including the knowledge of the EMOTIONALSTATE of the user yields an overall performance of MR/R = 60.6 % and thus contributes merely with another 0.1 % in contrast to the ASR set. We further have to bear in mind that hand-annotated emotions have been employed, implying a 100 % recognition rate in case of an automatic emotion recognition. An emotion recognizer itself is error-prone and a distinction of the emotional state of the user with the employed annotation scheme can be expected with approximately 60–80 % UAR, see Chap. 4 and e.g. Schmitt et al. (2010e), Polzehl et al. (2010c). The influence of emotion recognition on the IQ distinction can be considered as limited and is insofar not surprising as the occurrence of strong anger in the data is not dominant (5.0 %).

$\mathbf{IQ_{lab}}$ Performance

IQ_{lab} yields similar performance scores as IQ_{field}, which implies that the IQ framework results in comparable performance scores when it is applied on different system versions. Particularly the AUTO set achieves virtually the same performance. The EMOTIONALSTATE feature fails again and adds only another 0.4 %. A considerable performance gain can be achieved by including the USER group and by that context information about the respective user. The correlation between the label and the prediction even achieves $\rho = 0.894$, which depicts almost perfect correlation. However, it should be kept in mind that this information is not available in automated classification.

$\mathbf{US_{lab}}$ Performance

The US model achieves satisfying results, but yields significantly lower performance scores throughout all feature groups. For example, the AUTO set achieves only MR/R = 49.2 % compared to 58.9 % (IQ_{lab}) and 58.4 % (IQ_{field}). This can be attributed to the fact that each user seems to perceive an interaction differently and a general model that ties objective system events to subjective satisfaction scores is obviously more difficult to obtain. Again, the EMOTIONALSTATE only contributes with additional MR/R = 0.6 %.

5.6.5 Impact of Optimization Through Feature Selection

As can be seen, the clear advantage of using feature selection in the IQ and US task is the reduction of the feature space, which is considerably reduced when applying the selection process. The second benefit of feature selection, i.e., the increase of performance, can for example be observed in the relevant AUTO groups of IQ_{field} and IQ_{lab}. IQ_{field} yields MR/R = 61.6 % with 20 of 47 features, which is an increase of 3.2 %. IQ_{lab} achieves MR/R = 59.8 % and increases by 0.9 % by using only the 19 most relevant ones.

However, feature selection does not improve in all cases, which becomes clear when looking at the mean performance gain obtained through feature selection. This can presumably be attributed to the small size of the optimization set and the fact that the feature dimensionality still fits the available amount of data examples. Consequently, the Hughes effect does not seem to appear. Especially the small size

Table 5.7 Model performance after Leave-one-user-out (LOUO) validation on the training/test set. Depicted are the majority baseline (MB), if feature selection (FS) was involved, as well as MR/R, κ and ρ values

	IQ_{field}				IQ_{lab}				US_{lab}			
Input	FS	MR/R	κ	ρ	FS	MR/R	κ	ρ	FS	MR/R	κ	ρ
	MB	0.200	0.0	NA	MB	0.200	0.0	NA	MB	0.200	0.0	NA
DACT	no	26.9 %	0.0136	0.363	no	23.8 %	0.092	0.197	no	20.5 %	0.016	0.083
ASR	no	60.5 %	0.551	0.753	no	58.8 %	0.576	0.811	no	48.1 %	0.426	0.625
SLU	no	25.0 %	0.083	0.293	no	33.1 %	0.254	0.415	no	26.9 %	0.120	0.256
DM	no	42.9 %	0.334	0.653	no	48.7 %	0.471	0.772	no	38.1 %	0.318	0.619
AUTO	no	58.4 %	0.526	0.776	no	58.9 %	0.582	0.856	no	49.2 %	0.442	0.668
AUTO+EMO	no	60.6 %	0.549	0.785	no	59.3 %	0.587	0.856	no	49.8 %	0.443	0.669
AUTO+EMO+USER	–	–	–	–	no	64.1 %	0.640	0.888	no	58.1 %	0.523	0.741
ALL	no	61.9 %	0.559	0.800	no	64.7 %	0.648	0.894	no	59.2 %	0.529	0.742
DACT	–	–	–	–	–	–	–	–	–	–	–	–
ASR	13/25	59.8 %	0.545	0.730	12/25	59.2 %	0.580	0.816	13/25	48.4 %	0.434	0.623
SLU	4/5	25.0 %	0.083	0.293	5/5	33.1 %	0.254	0.415	4/5	26.3 %	0.117	0.259
DM	10/17	43.6 %	0.338	0.649	4/17	47.6 %	0.458	0.748	5/17	37.9 %	0.316	0.589
AUTO	20/47	61.6 %	0.563	0.786	14/47	59.8 %	0.598	0.839	19/47	49.2 %	0.437	0.647
AUTO+EMO	31/48	60.4 %	0.545	0.785	14/48	59.8 %	0.598	0.839	19/48	49.2 %	0.437	0.647
AUTO+EMO+USER	–	–	–	–	13/51	59.8 %	0.598	0.839	29/51	60.3 %	0.555	0.758
ALL	23/52	62.5 %	0.575	0.795	13 / 56	59.8 %	0.598	0.839	29/56	60.3 %	0.555	0.758
Mean FS gain		+0.6 %	+0.008	−0.004		−1.2 %	−0.011	−0.022		+0.3 %	+0.007	−0.006

The first half comprises results when all features of a group are employed. The second half contains results after feature selection on the optimization set. The feature number with best performing scores were selected to cross-validate the training/test set with LOUO classification ((x/y) = where x is the number of features used from all y available features). Mean FS Gain depicts the mean improvement that was obtained through feature selection

Table 5.8 All classification results for cross-target prediction

Feature set	Test	Train	MR/R (%)	κ	ρ
AUTO	US_{lab}	IQ_{field}	32.8	0.191	0.487
AUTO	US_{lab}	IQ_{lab}	39.1	0.363	0.667
AUTO	IQ_{lab}	US_{lab}	24.8	0.161	0.435
AUTO	IQ_{lab}	IQ_{field}	39.7	0.311	0.647
AUTO	IQ_{field}	IQ_{lab}	38.2	0.268	0.696

For evaluation the AUTO parameters have been used and all available data, i.e., train/test and optimization set, have been respectively employed for training and testing

of the optimization set in the smaller LG-LAB corpus and the resulting suboptimal feature selection seems to harm the performance.

5.6.6 Cross-Target Prediction and Portability

A cross-target evaluation is introduced to give some indication of the portability of the models. Results are depicted in Table 5.8. In general it is visible that the values achieve lower scores than when training and testing has been conducted on the same target variable as in Table 5.7. According to the table, the actual satisfaction of a lab user can be in average predicted with MR/R = 32.8 % if the model has been trained on expert annotations from the field corpus. While this seems low at first sight, it should be reminded that the employed dataset for training originates from a different system version of Let's Go. The low performance of IQ_{lab} when trained on US_{lab} (MR/R = 24.8 %) indicates that user satisfaction scores do not seem to qualify for creating robust models due to the high influence of subjectivity. Interchanging training and testing with IQ_{lab} and IQ_{field} yields a constant performance of approx. 39 %.

5.6.7 Causalities and Correlations Between Interaction Parameters and IQ/US.

While we have successfully classified IQ and US in the previous sections and analyzed, which parameters are most relevant for *classification*, we have not yet considered, if a linear dependency between input- and target variable exists and which factors eventually influence the user satisfaction.

The PARADISE approach presumes a linear relationship between input variables—quantifying the dialog—and the target variable US, the user satisfaction. Assuming linearity, linear regression models can be applied that allow inferences such as "The longer the average dialog duration in an SDS, the lower the average user satisfaction."

Although linear regression models yield lower performance values on complex problems than non-linear models (as e.g. SVMs), see Schmitt et al. (2011a), they may allow visualization of dependencies between dependent variables (the features) and independent variables (the respective IQ/US scores). A linear regression model of IQ respectively US is calculated as follows:

$$\mathcal{N}(\mathrm{IQ/US}) = \sum_{i=1}^{n} w_i \times \mathcal{N}(p_i), \tag{5.8}$$

where w_i is the weight for the interaction parameters p_i, and $\mathcal{N}$ the z-score normalization function. $\mathcal{N}$ normalizes the input variables to a mean value of zero and a standard deviation of one. This eliminates the varying scales of the input variables.

Linear regression models of $\mathrm{US}_{\mathrm{lab}}$, $\mathrm{IQ}_{\mathrm{lab}}$ and $\mathrm{IQ}_{\mathrm{field}}$ with the top 15 input variables are depicted in Table 5.9. They have been obtained using all data and the respective *ALL* feature sets. Nominal input variables such as UTTERANCE, ACTIVITYNAME and INTERPRETATION have been removed, since they can not be mapped to a suitable continuous scale. Moreover, other nominal variables that can be brought into a linear ordinal ranking have been mapped to an ordinal scale, such as EMOTIONALSTATE {garbage and non-angry = 1, slightly-angry = 2, very-angry = 3}.

The linear regression functions reveal a central property of the IQ and US schemes as we have previously observed in the IGR ranking in Table 5.6. The duration of the dialog plays a central role and has the most relevant impact on the scores (TURNNUMBER, #SYSTEMTURNS, #USERTURNS). Similarly, variables that are usually increased with progressing dialog can be found among the highest ranked coefficients, such as #ASRSUCCESS, #REPROMPTS and #TIME- OUTPROMPTS. Taking a closer look at the US model shows that the US score is mostly dominated by ASR performance parameters, such as %TIMEOUTS_ASRREJ, %ASRREJECTIONS, %ASRREJECTIONS and %REPROMPTS. This indicates that users strongly tie their subjective satisfaction on the system's understanding capability. Interestingly enough, the emotional state does not occur among the top 15 coefficients in any of the models. A straight linear dependency between (independent) input and the (dependent) target variables does not seem to exist. This becomes obvious by considering the $\mathrm{US}_{\mathrm{lab}}$ model details. It seems natural that with an increasing %TIMEOUTS_ASRREJ rate, US decreases. However, an increasing %ASRREJECTIONS rate would increase US, according to the model. Similar examples can be found when looking at $\mathrm{IQ}_{\mathrm{lab}}$ and $\mathrm{IQ}_{\mathrm{field}}$.

To shed further light upon the dependency of user satisfaction and the interaction parameters we analyze their correlations, which are depicted in Table 5.10. It is obvious that it is less probable to observe low US scores in the beginning of a dialog. Low US is assumed to occur only after a number of negative interaction patterns. It is thus not surprising that variables that are increased with progressing dialog and that contain absolute values highly negatively correlate with US, such as #USERTURNS, #ASRSUCCESS, TURNNUMBER, #SYSTEMTURNS and #REPROMPTS. In this context, the variables on a percentage basis are more illuminating, such as %REPROMPTS and

Table 5.9 Linear regression models for US_{lab}, IQ_{lab} and IQ_{field} with first 15 most relevant coefficients

$US_{lab} =$	$IQ_{lab} =$	$IQ_{field} =$
$-1.2 \cdot \mathcal{N}$(%TimeOuts_ASRRej)	$+1.3 \cdot \mathcal{N}$(%TimeOuts_ASRRej)	$-1.1 \cdot \mathcal{N}$(#ASRSuccess)
$+0.9 \cdot \mathcal{N}$(%Time- OutPrompts)	$-1.0 \cdot \mathcal{N}$(%Time- OutPrompts)	$+1.0 \cdot \mathcal{N}$(#RePrompts)
$+0.7 \cdot \mathcal{N}$(%ASRRejections)	$-0.8 \cdot \mathcal{N}$(%ASRRejections)	$-1.0 \cdot \mathcal{N}$(#UserTurns)
$-0.1 \cdot \mathcal{N}$(%RePrompts)	$-0.6 \cdot \mathcal{N}$(#ASRSuccess)	$+0.6 \cdot \mathcal{N}$(TurnNumber)
$-0.1 \cdot \mathcal{N}$(#ASRSuccess)	$-0.6 \cdot \mathcal{N}$(#UserTurns)	$+0.6 \cdot \mathcal{N}$(#SystemTurns)
$-0.1 \cdot \mathcal{N}$(#UserTurns)	$-0.3 \cdot \mathcal{N}$(%RePrompts)	$-0.6 \cdot \mathcal{N}$(#SystemQuestions)
$+0.1 \cdot \mathcal{N}$(#RePrompts)	$+0.3 \cdot \mathcal{N}$(#RePrompts)	$-0.5 \cdot \mathcal{N}$(%RePrompts)
$-0.1 \cdot \mathcal{N}$(RoleIndex)	$+0.2 \cdot \mathcal{N}$(TurnNumber)	$-0.3 \cdot \mathcal{N}$(#TimeOuts_ASRRej)
$+0.1 \cdot \mathcal{N}$(#Time- OutPrompts)	$+0.2 \cdot \mathcal{N}$(#SystemTurns)	$+0.2 \cdot \mathcal{N}$(#Barge- Ins)
$-0.1 \cdot \mathcal{N}$(TurnNumber)	$-0.1 \cdot \mathcal{N}$(#ASRRejections)	$-0.2 \cdot \mathcal{N}$({#}ASRSuccess)
$-0.1 \cdot \mathcal{N}$(#SystemTurns)	$-0.1 \cdot \mathcal{N}$(#TimeOuts_ASRRej)	$+0.2 \cdot \mathcal{N}$(MeanASRConfidence)
$+0.1 \cdot \mathcal{N}$(ComputerDailyMin)	$+0.1 \cdot \mathcal{N}$(#SystemQuestions)	$-0.1 \cdot \mathcal{N}$(#ASRRejections)
$-0.1 \cdot \mathcal{N}$(#HelpRequests)	$+0.1 \cdot \mathcal{N}$(TS)	$+0.1 \cdot \mathcal{N}$(%ASRRejections)
$+0.1 \cdot \mathcal{N}$(userId)	$-0.1 \cdot \mathcal{N}$({#}ASRSuccess)	$+0.1 \cdot \mathcal{N}$(% Success)
$-0.1 \cdot \mathcal{N}$(#ASRRejections)	$-0.1 \cdot \mathcal{N}$(%ASRSuccess)	$-0.1 \cdot \mathcal{N}$(#Time- OutPrompt)
[..]	[..]	[..]
(5.9)	(5.10)	(5.11)

A linear dependency between the input variables and the target variables does not seem to exist

Table 5.10 Input variables with positive or negative correlation to US

Parameter	Spearman's ρ
#USERTURNS	−0.64
#ASRSUCCESS	−0.64
TURNNUMBER	−0.62
#SYSTEMTURNS	−0.62
#REPROMPTS	−0.60
#SYSTEMQUESTIONS	−0.51
%REPROMPTS	−0.48
#TIMEOUTS_ASRREJ	−0.42
#ASRREJECTIONS	−0.36
EMOTIONALSTATE	−0.31
#TIMEOUTS_ASRREJ	−0.31
%ASRREJECTIONS	−0.28
{#}ASRSUCCESS	−0.28
%ASRSUCCESS	+0.26
%TIMEOUTS_ASRREJ	−0.25
%TIME- OUTPROMPTS	−0.22
{#}REPROMPTS	−0.22
ROLEINDEX	−0.15
{#}TIMEOUTS_ASRREJ	−0.14
WPUT	−0.12
UTD	−0.11
ASRCONFIDENCE	−0.11
REPROMPT?	−0.11
{#}TIME- OUTPROMPTS	−0.10
{#}ASRREJECTIONS	−0.10
SDSATTITUDE	+0.09
COMPUTERDAILYMIN	+0.08
{#}SYSTEMQUESTIONS	−0.07
MEAN ASRCONFIDENCE	+0.06
WPST	+0.05

All correlations are highly statistical significant

%ASRREJECTIONS (negative sign), as well as the variables affecting a specific time frame in the dialog, e.g. {#}ASRSUCCESS (positive sign). Marginal, but interesting correlations can be observed on variables that affect the expertise of the users: a positive attitude towards SDS (SDSATTITUDE) and the daily computer usage time (COMPUTERDAILYMIN) tend to positively influence US. According to the (weak) negative correlation of WPUT and UTD we can conclude that dissatisfied users tend to speak more and longer to the system than satisfied users.

5.7 Summary and Discussion

We have presented a generic framework for modeling and predicting IQ in ongoing system- user interaction in SDS, which is applicable in field and lab scenarios. In contrast to plain emotion recognition, the model allows to spot poor quality during an interaction even when users do not show obvious emotional affects. The paradigm allows to estimate SDS performance on the exchange level and by that allows online monitoring of the interaction in an SDS at arbitrary points in time. In case the model estimates low IQ, the dialog manager may trigger adaptivity steps. It can be further employed in offline scenarios by system designers for spotting critical dialog situations and for uncovering poor system design. The approach follows roughly the idea of PARADISE (Walker et al. 2000), but differs mainly in four points:

- PARADISE aims to determine overall system performance, i.e., a kind of usability score, for an entire system. In contrast, the IQ framework aims to determine the quality of a specific interaction between the system and the user.
- Thereby, the modeling of the interaction by means of interaction parameters is performed on the exchange level and not on the dialog level. They serve as input variables for the model.
- Furthermore, the main target variable is an objective expert-annotated score (IQ) and not a subjective user score.
- The classification mainly resides on non-linear support vector models instead of linear regression models.

The paradigm has been evaluated on data from both field and lab scenarios of the CMU Let's Go bus information system. While the former originates from real users interacting with the system in real tasks, the latter has been collected in the course of this study with 46 users. The study at hand showed that subjective user satisfaction highly correlates with the objective IQ expert annotations. For this comparison, user satisfaction scores have been obtained in the lab study, while expert annotations on the same data resulted in a more objective view on the quality. The Interaction Quality paradigm thus resolves the problem of tracking real user satisfaction in field environments and instead suggests that expert annotations are more suitable to model the performance of an interaction than US is. In the following classification experiments it could be further shown that IQ predicts better than US. The classification models have been created with SVMs and each single feature group was evaluated along with the respective target variables. The most relevant predictors for IQ and US are related to ASR performance. A linear dependency between the input variables and the target variables seems less likely, which could be demonstrated on the basis of PARADISE-style linear regression models.

Portability

The advantage of such an expert-based approach is the straightforward reproducibility without the need for user tests, which are difficult to implement in the field. The guidelines and the use of multiple raters assure that similar results can be obtained in different system versions. This could be shown by applying the paradigm on two different versions of the Let's Go Bus information system. We assume that, the paradigm yields similar performance values when applied on other domains, since the presented interaction parameter set and the guidelines are domain independent. A proof of this assumption still has to be adduced. We further assume that porting the paradigm on other systems would require an annotation process for the respective system, since the parameters themselves are domain-independent. Some parameter *values*, however, are system-dependent, such as UTTERANCE that heavily depends on the system's grammar.

US Study

The user satisfaction study is to our knowledge the first one to measure user satisfaction at arbitrary points with a real spoken dialog system, i.e., the data has not been captured in a Wizard-of-Oz scenario as in Engelbrecht et al. (2009). In contrast to Engelbrecht et al. it further measures satisfaction without the need for interrupting the dialog. Although initially assumed, the users did not have difficulties in adjusting their satisfaction during the interaction, since in only 6 % of the dialogs the users indicated that they missed to mirror their satisfaction with the input device. However, it cannot be entirely excluded that some users sometimes failed to adjust their US score and did not indicate this. It should be put up for discussion that the users in the study are not citizens of Pittsburgh and are thus not familiar with the transportation system and the geography of the region, where the instrumented Let's Go Bus Information System is designed for. We assume that this has only a minor impact on the quality of the data. Similarly, the effect of the users being non-natives can presumably be neglected since a large proportion of the real Let's Go users are non-natives themselves (Raux and Eskenazi 2004) and the language capabilities of the participants can be considered as high. A major difference of field and lab users concerns the endurance of the two groups: notwithstanding that the lab users have been encouraged to hang up if they would do so in a field scenario, many users showed a remarkable patience to finish the task. Notably unsolvable scenarios, e.g. where the system stated that no schedule is available, seemed to challenge the users. While the lab users perpetually tried different ways to solve the task, users in the field gave up much more quickly.

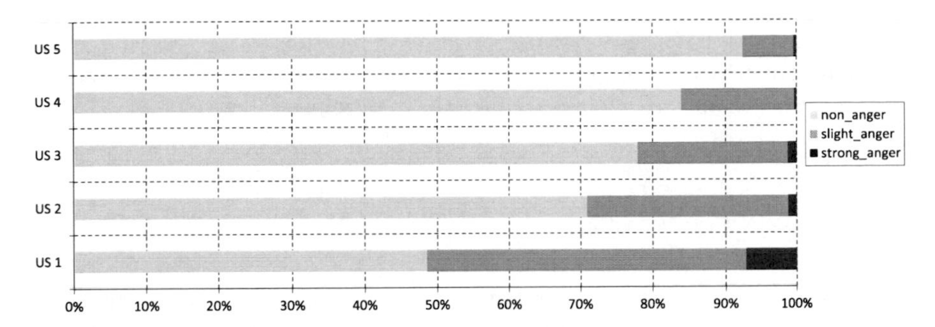

Fig. 5.9 Subjective user satisfaction (US_{lab}) versus annotated user anger

Expert Ratings Versus User Satisfaction

Obviously, the performance of an SDS rated by an expert annotator and the subjective impression of users do highly correlate. It is further interesting to note that the correlation among the expert annotations ($\rho = 0.72$; $\rho = 0.67$) is only slightly higher than the correlation between experts and subjective user impressions ($\rho = 0.66$). When considering agreement, the values achieve comparable scores among the single experts ($\kappa = 0.31$; $\kappa = 0.29$) as the agreement between experts and users ($\kappa = 0.31$). This implies that raters and users experience the dialog interaction similarly, notwithstanding their actual involvement in the task.

Negative Emotions

For the first time, the study furthermore brought to light how users perceive an interaction and how they behave emotionally. Obviously, there exists a dependency between objectively observed user anger and subjective user satisfaction, see Fig. 5.9.

A large proportion of users that assigned the scores "1" and "2" expressed their dissatisfaction emotionally. US and EMOTIONALSTATE correlate with $\rho = -0.31$, i.e., the more dissatisfied the user was according to the rating, the more he expressed his dissatisfaction using emotions. The reverse conclusion is: a large proportion of the users do not react emotionally when being dissatisfied (48 % at a score of "1"). This implies that the use of emotion recognition for spotting dissatisfied users *alone* is not sufficient and depicts another motivation for IQ models, i.e., models as presented in this work are of need to spot a larger proportion of dissatisfied users. Interestingly enough, the emotion information has no impact on the classification performance. A model that is endowed with the EMOTIONALSTATE feature derived from the user's emotional state does not yield considerably higher performance scores than models trained without emotions. Furthermore, even the performance of predicting low scores could not achieve much better with the EMOTIONALSTATE feature than without. Moreover, the absence of EMOTIONALSTATE in the top 10 features in the IGR list and the top 15 coefficients of the linear regression models indicates that emotion recognition in this specific scenario is only of minor help.

Classification Performance

The presented models yield very satisfying performance scores. The employed main performance metric MR/R acts as discriminative and not as regressive performance metric and punishes rather merciless even close misclassifications. The discrimination of five classes by using only automatic features yields 61.6, 59.8, and 49.8 % respectively for IQ_{field}, IQ_{lab} and US_{lab}. Considering the correlation of the labels and the model predictions we respectively obtain $\rho = 0.786$, $\rho = 0.839$ and $\rho = 0.668$ for IQ_{field}, IQ_{lab} and US_{lab}, which altogether can be considered as highly correlating. Taking into account further non-automatic features yields even higher scores. The good performance of the model becomes clear when the κ and ρ scores obtained are compared to the inter-rater agreements. For example, the IQ_{field} model and the target label IQ_{field} obtain a higher agreement score with $\kappa = 0.563$ than the three single raters among each other with $\kappa = 0.31$ (see again Table 5.5). For ρ the values are 0.786 versus 0.72. Even when analyzing the agreement of the single raters with the target variable IQ_{field}, which in average achieved an agreement of $\kappa = 0.608$ and a correlation of $\rho = 0.874$ (see again Table 5.3), the model can still keep pace with $\kappa = 0.563$ and $\rho = 0.786$. For IQ_{lab} and US_{lab} similar observations can be made, which are not further discussed here.

Static Modeling

As IQ and US scores depict time series within a dialog, other statistical modeling techniques than the proposed one may be taken into consideration in future work. In the presented approach IQ and US are modeled with a static feature vector. Prior information occurring temporally before the current dialog step are included with the context parameters (with suffixes '{MEAN}' and '{#}) and with dialog-wide mean values and running tallies of important system events (suffixes '%' and '#'). Nevertheless, a higher resolution of the temporal characteristics of the previous interaction may be achieved by modeling the time series with Hidden Markov Models or conditional random fields (CRF), which are specialized on time series prediction and which might achieve slightly higher accuracies.

Chapter 6
Statistically Modeling and Predicting Task Success

The PARADISE approach postulates that user satisfaction is maximized when task success rates are maximized and dialog costs minimized. While this assumption may hold for the evaluation of dialog systems and completed dialogs, task success cannot be taken into account when estimating user satisfaction *during* an interaction as the task success is not ascertainable when a dialog is not finished. The previously presented IQ approach is thus entirely based on efficiency and qualitative measures and omits any information (i.e. features) about task success.

Notwithstanding this, the estimation of a successful dialog outcome is a relevant function for online monitoring SDS interactions and will enable further adaptivity in future systems. As SDS are neither by design nor technologically omnipotent and infallible, strategies are required that recognize the risk of task failure. With the information that a task between the system and the user ends in failure, an SDS may switch to a fallback strategy for example by changing the prompt strategy or by seeking human assistance in form of professional human agents.

This latter escalation to human assistance is particularly an option for SDS deployed in call-center environments. There, task success prediction also shows commercial character. In call-centers, SDS are used to reduce the proportion of human operators in telephone conversations, most importantly to reduce handling costs. The average handling time for SDS automation also rises along with the increasing complexity of the task an SDS has to resolve.[1] Hence, it has become common practice to let the customers start with automated systems handling the standard cases and to dispatch the call to a human operator when a number of misrecognitions occur or the call reason has been identified to be out of the system's scope. If the system fails, e.g. when the call is out of scope, a human operator usually takes over the dialog system's role driving the overall expense for this call up. In such a case, the expense of a non-automated call is the automated portion's handling fee plus the human operator's fee. Both of them directly depend on the respective handling time.

---

[1] As for the LevelOne Automated Troubleshooters, a substantial amount of calls last more than 10–20 min.

A. Schmitt and W. Minker, *Towards Adaptive Spoken Dialog Systems*,
DOI: 10.1007/978-1-4614-4593-7_6,

Taking this preamble into account, it seems obvious that calls that do not have any potential to be automated by the dialog system should be escalated to a human agent as soon as possible in order to reduce the automation handling fee. This also lies in the interest of the customer whose time is otherwise wasted with a system that does not solve his problem.

Some systems deployed in the field use static rules taking into account the number of misrecognitions, time-out events, and out-of-scope inputs as indicator for task failure. Once a certain threshold of these symptoms is exceeded, the call is escalated to an operator. While the occurrence of the mentioned events may have a considerable correlation with task success (i.e. whether the call will be automated or not), there are many other factors providing additional information on the expected dialog outcome such as the current dialog step the user is in, the duration of the dialog so far, the barge-in behavior, etc. Predicting task success is therefore better solved on a statistical basis, allowing a more flexible and robust approach.

In the following, such a statistical modeling technique is presented that allows to estimate task success for ongoing human–computer dialogs. Similarly as in the estimation of Interaction Quality, a large set of input variables is employed to predict the probability of task success.

In Sect. 6.1 we will first present an approach that we will refer to as 'linear modeling' approach, as it takes linearly all information captured from the previous interaction into account to estimate task success. As this leads to an unmanageable amount of data with increasing dialog progress, we define in Sect. 6.2 an improved approach that reduces the number of required parameters while keeping the performance on a comparable level. Under the assumption that the current dialog step in a dialog flow has a special importance for predicting task success, we defined an SRI-based statistical model and tie system actions to outcome classes in Sect. 6.3.

As we will see, the prediction of task success comes with a high uncertainty that has to be dealt with. The model's certainty about task failure needs to be sufficiently high before it recommends aborting an interaction by escalating to a human agent. This problem is addressed in Sect. 6.4.

The presented approaches are evaluated in Sect. 6.5 and finally we conclude with a summary and discussion in Sect. 6.6.

6.1 Linear Modeling

We follow a formal definition to model the prediction of task success in SDS interactions. Let the exchange

$$\varepsilon = \{\pi_1, \ldots, \pi_n\} \tag{6.1}$$

be a dialog exchange represented by automatic interaction parameters π_i

$$\pi \in \{\text{ASR}, \text{SLU}, \text{DM}\}, \tag{6.2}$$

where ASR, SLU, and DM represent the interaction parameters introduced in Sect. 3.2.1, excluding the window- and dialog-level parameters.[2] Further let

$$\sigma_\lambda = \{\varepsilon_1, \ldots, \varepsilon_\lambda\} \tag{6.3}$$

be a sequence of such dialog exchanges ε_i representing the interaction of a dialog δ from exchange 1 up to exchange λ. Further let

$$\theta_\lambda = \{\{\sigma_{\lambda,1}, \omega_i\}, \ldots, \{\sigma_{\lambda,j}, \omega_i\}\} \tag{6.4}$$

be a training set of size j consisting of exchange sequences σ_i leading to one of the outcomes $\omega_i \in \Omega = \{\omega_1, \ldots, \omega_n\}$. Then a statistical classifier can be trained with the training set θ_λ resulting in a model μ_λ. To predict the outcome of an ongoing dialog with the current length λ we determine σ_λ and estimate

$$\hat{\omega} = \arg\max_{\omega} P(\omega|\sigma_\lambda, \mu_\lambda). \tag{6.5}$$

It should be noted that with the formalism described above we are only capable of predicting the outcome at a very specific point in the conversation, namely at exchange ε_λ. From now on, we will refer to this exchange, where we classify the dialog outcome, the classification point γ. A system Π capable of predicting the outcome in a dialog system at several exchanges $1, \ldots, k$, i.e. at several classification points γ_k consists of a set of models $\{\mu_1, \ldots, \mu_k\}$. For achieving this, a separate model has to be trained for each possible classification point γ_i. In other words, task success can be predicted using the dense information representing a dialog conversation up to a certain point in the conversation. The model allows a comparison of the currently observed pattern with patterns that have previously been observed in other dialogs. At each single classification point, the system draws on the respective model in order to determine the outcome of the dialog.

The approach is illustrated in Fig. 6.1.

It should be noted that the classification algorithm that is employed to determine the outcome $\hat{\omega}$ is finally not of major importance as long as it belongs to the family of discriminative classifiers. Possible classifiers are e.g. artificial neural networks (ANN), k-nearest neighbor (kNN), support vector machines (SVM), Decision Trees or Rule Learners, such as RIPPER (Cohen and Singer 1999).

6.2 Window Modeling

The approach described above entails a central disadvantage: the later the classification point γ, the larger the feature vector that is used to represent the dialog course and that is ultimately used to train the model. This approach may lead to an unman-

[2] I.e. parameters with prefixes {#}, #, {Mean}, Mean.

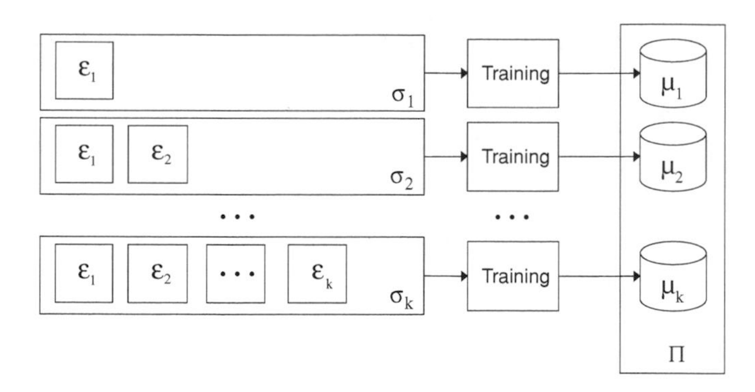

Fig. 6.1 Linear modeling: the system Π for predicting task success consists of k models μ_k. Each model is trained with interaction parameters from all previous exchanges $\varepsilon_{1 \ldots \lambda}$. The size of the feature vector for training increases linearly with progressing k (cf. Walker et al. 2002; Kim 2007; Schmitt et al. 2008a)

ageable vector dimension and may finally end up causing the Hughes effect (Hughes 1968).

It can be assumed that the outcome of a dialog ultimately depends less on the entire interaction than on a certain window prior to the classification point γ. To be more precise, when estimating the outcome of a dialog at e.g. classification point $\gamma = 15$, a certain detail from a very early point of the dialog, say from exchange $\varepsilon = 3$, may have no effect on the outcome. For example, the information about a "no-match event" (the user utterance could not be recognized) from an early exchange is not as meaningful as a "no-match event" occurring in the recent context. As a consequence it might be sufficient to model the interaction of the recent turns in detail, and project central characteristics of the remaining interaction into summarizing variables. Such characteristics can entail e.g. the number of no-match and no-input events, the number of help requests, the number of operator requests etc. Characteristics affecting the immediate *context* κ of ε_λ are defined as

$$\kappa_\lambda = \{\pi_1, \ldots, \pi_n\}, \tag{6.6}$$

where

$$\pi_i \in \{\{\#\}\text{ASR}, \{\%\}\text{ASR}, \{\#\}\text{SLU}, \{\%\}\text{SLU}, \{\#\}\text{DM}, \{\%\}\text{DM}\}, \tag{6.7}$$

and characteristics quantifying the *total* interaction τ up to ε_λ are defined as

$$\tau_\lambda = \{\pi_1, \ldots, \pi_n\}, \tag{6.8}$$

where

$$\pi_i \in \{\#\text{ASR}, \%\text{ASR}, \#\text{SLU}, \%\text{SLU}, \#\text{DM}, \%\text{DM}\}. \tag{6.9}$$

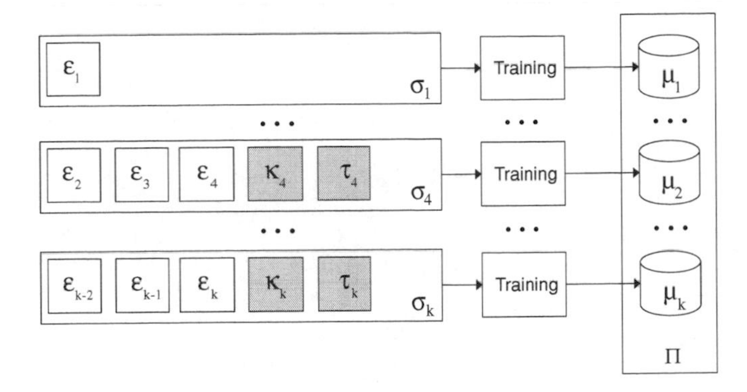

Fig. 6.2 Window modeling: only the last n turns (here $n = 3$) are included for modeling task success. Interaction going back further is captured less detailed with summarizing parameters κ for the immediate context, and τ for the entire interaction

Consequently, we use a window of n *recent exchanges* (an n-gram) to represent the interaction and not—as in the previous linear approach—the *entire exchanges*. Such an n-gram is then defined as follows:

$$\sigma_\lambda = \{\varepsilon_{\lambda-n}, \ldots, \varepsilon_{\lambda-1}, \varepsilon_\lambda, \kappa_\lambda, \tau_\lambda\}. \tag{6.10}$$

The approach is illustrated in Fig. 6.2.

Simply speaking, the decision whether a dialog outcome will be successful or not is now based on observations from the previous n exchanges (where the resolution is high) plus certain performance scores that emphasize the interaction of the immediate context, as well as performance scores from the total interaction (where the resolution is low). Estimating the outcome class is accomplished by applying Eq. 6.5.

6.3 SRI and Salience

It can be assumed that it is not only the interaction between the system and the user that decides whether a task will be completed, i.e. the barge-in behavior, the performance of the ASR etc., but also the difficulty of the task and particularly the current dialog step the user is in. In the interaction parameter set, this information is contained in the parameter ACTIVITYNAME. Figure 6.3 depicts the dialog flow structure of the SpeechCycle Broadband Agent system derived from the SC-BAC database. Many telephone-based SDS follow such a tree structure. In the first step of each dialog in the Broadband Agent system, the customer is welcomed and some details are looked up from the database depending on the caller's phone number. In the first question the customer is asked whether his problem is related to Internet connectivity. Here, the major problem in the broadband troubleshooting domain is separated from the less

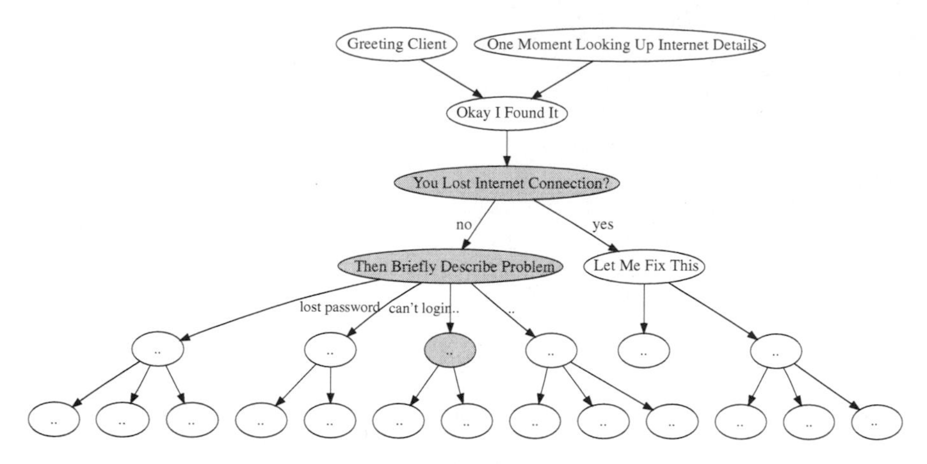

Fig. 6.3 Typical structure of a technical support automated agent and a majority of IVR systems using the example of SpeechCycle LevelOne Broadband Agent. It should be noted that each node stands for an ACTIVITYNAME. The highlighted nodes depict a trigram of ACTIVITYNAMES

frequently occurring other problems. In case the user is calling because of another issue than Internet connectivity, he is asked to describe his problem in the next step. As we can see, the tree broadens with each level and the position in the branches mirror the problem and the problem solution progress. From the tree structure it seems obvious that a sequence of ACTIVITYNAMES already indicates the outcome of the dialog.

In Sect. 4.1.2 we have seen that certain words occur more frequently with a specific emotion class than others. For example the words "operator" and "wrong" co-occurred more often with the class ANGER than with the class NONANGER. On the other hand, some words appeared more frequently with the NONANGER class, such as "thanks" and "right", and other words did not even carry a specific emotional coloring. This "emotional salience" approach (Lee and Narayanan 2005) can be adapted to meet the requirements for predicting the outcome of a dialog: certain sequences of activities might more frequently co-occur with an outcome FAIL than with the outcome SUCCESS and vice-versa.

We first denote an n-gram of ACTIVITYNAMES as

$$\alpha_k = \{\text{ACTIVITYNAME}_{k-n}, \ldots, \text{ACTIVITYNAME}_{k-1}, \text{ACTIVITYNAME}_k\} \quad (6.11)$$

and the set of outcome classes as $\Omega = \{\omega_1, \omega_2, \ldots, \omega_m\}$.[3] The SRI contained in a single activity n-gram α_k in relation to an outcome class ω_i is then given by

$$i(\alpha_k, \omega_i) = \log \frac{P(\omega_i|\alpha_k)}{P(\omega_i)}. \quad (6.12)$$

[3] For example $m = 3$, $\Omega = \{\text{SUCCESS}, \text{PARTIALSUCCESS}, \text{FAILURE}\}$.

Table 6.1 Most salient trigrams of ActivityNames for SC-BAC

Nr.	Trigram description	Salience	Class
1	A live agent is offered to the user and user accepts	2.78	FAILURE
...			
234	User decides to stay with the system after system prompts "You want to stay with me or finish with an operator?"	1.62	SUCCESS
...			
563	System has identified that the Internet problem is tied to wrong operating system version	0.29	PARTIALSUCCESS
...			
889	Check modem and determine DNIS	0.01	FAILURE

Here, $P(\omega_i|\alpha_k)$ denotes the posterior conditional probability that the activity n-gram α_k implies outcome ω_i, i.e. i expresses the dependency between an activity and a specific outcome class. $P(\omega_i)$ is the prior probability of this outcome class. Consequently, if $P(\omega_i|\alpha_k) > P(\omega_i)$, then $i(\alpha_k, \omega_i)$ is positive implying that this outcome ω_i is more likely. If $P(\omega_i|\alpha_k) < P(\omega_i)$, then $i(\alpha_k, \omega_i)$ is negative, which makes the specific outcome ω_i less likely.

The salience of an Activity n-gram can then be defined as

$$\mathrm{sal}(\alpha_k) = I(\Omega; \alpha_k) = \sum_{j=1}^{m} P(\omega_j|\alpha_k) i(\alpha_k, \omega_j). \tag{6.13}$$

The higher the salience value, the stronger the correlation with the outcome classes. The Broadband Agent system contains 889 different activities resulting in 13,550 unique activity trigrams. Table 6.1 depicts the three top-most salient trigrams of the SC-BAC database of each of the three outcome classes. The fourth entry depicts a trigram with a very low salience. As this trigram usually appears in the beginning of a dialog, it does not contain appreciable discriminative power. It should be noted that the differences between the salience values are rather high in this example, which can be attributed to the very skew distribution of the outcome classes in the SC-BAC corpus. It seems evident that a call where trigram number 1 is observed virtually always ends as FAILURE dialog, as the system has already reached a state where the user refused to continue with the system. In contrast to this, trigram number 234 is merely an indicator for a successful dialog, as the user seems to show cooperative behavior. This is certainly no guarantee for a successful outcome, but a strong hint.

From the presented salience approach, we include the SRI i of all possible outcome classes Ω into task success modeling and extend Eq. 6.10, i.e. the feature vector, by ξ

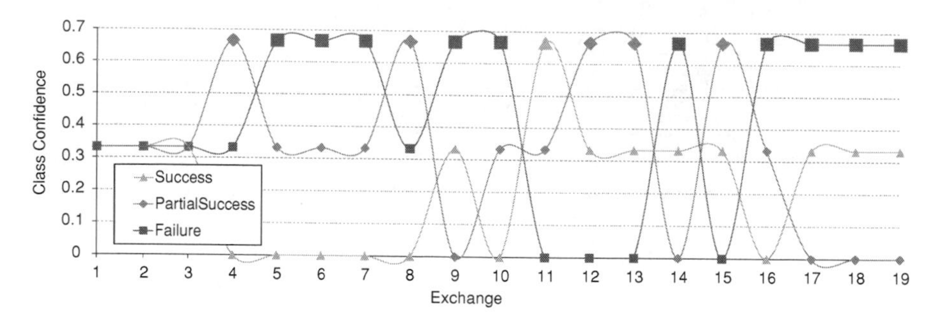

Fig. 6.4 Model chart derived from the Witchcraft Workbench (Schmitt et al. 2010a, b) when a task success prediction model is applied on a specific example dialog (Schmitt et al. 2010f). The *lines* symbolize the confidence scores of the classifier at each system–user exchange in the dialog for all three possible dialog outcomes. The *upmost markers* at confidence level 0.66 symbolize the classifier's hypothesis. The dialog depicted here belongs to the class FAILURE

$$\sigma_\lambda = \{\varepsilon_{\lambda-n}, \ldots, \varepsilon_{\lambda-1}, \varepsilon_\lambda, \kappa_\lambda\, \tau_\lambda, \xi_\lambda\}, \tag{6.14}$$

where $\xi = \{i(\alpha_\lambda, \omega_1), \ldots, i(\alpha_\lambda, \omega_k)\}$.

6.4 Coping with Model Uncertainty

During the past years dealing with the prediction of task success the question was risen how a statistical classifier would 'behave' when classifying a specific dialog at arbitrary points in an ongoing interaction. This was a central motivator for the Witchcraft Workbench, cf. Sect. 3.3 and Schmitt et al. (2010a, b). It displays predictions of discriminative and regressive classifiers jointly with a dialog flow. It enables a view on the classifier's predictions at arbitrary points in the dialog. When applying the task success prediction model described in Sect. 6.2 on specific dialogs, Witchcraft creates charts for the estimated dialog outcome as depicted in the example in Fig. 6.4.

The applied classifier discriminates between three possible dialog outcomes: SUCCESS, PARTIALSUCCESS, and FAILURE. The chart shows the classifier's confidence of observing a dialog, i.e. call, that belongs to one of the three classes. The bold topmost dots at the 66.6 %-confidence line represent the classifier's hypothesis, i.e. the prediction.

N-Tuples as Decision Rule

A number of findings can be observed when applying the model on a large number of dialogs from the SC-BAC data set and analyzing the outcome charts. First, a majority of the dialogs exhibit such patterns, where the classifier varied strongly between the adjacent system–user exchanges. Second, the classifier obviously

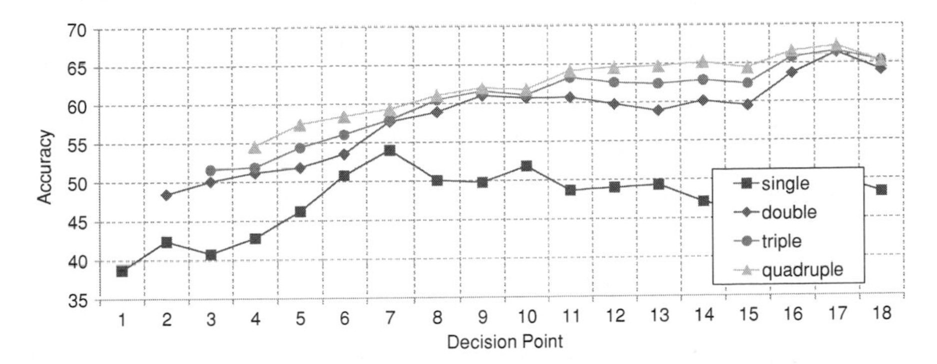

Fig. 6.5 Prediction accuracies of the model when double, triple, and quadruples are taken as basis of decision-making compared to a single prediction (cf. Schmitt et al. 2010f)

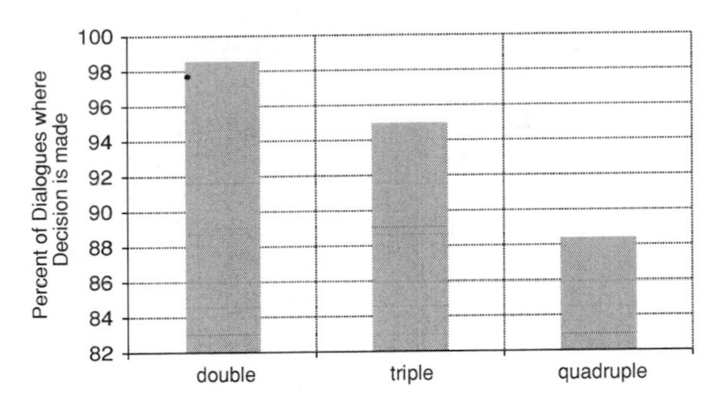

Fig. 6.6 Percentage of dialogs where a decision is taken

produces sequences of same predictions. In this chart the class FAILURE is predicted several times in a row at exchanges 5, 6, 7 and 9, 10 as well as 16, 17, 18, 19. When sequences like these occurred, they have been frequently correct, such as in this example, where the call indeed belongs to the class FAILURE.

It seems obvious to postpone any decision that causes far-reaching consequences, such as whether to escalate the call or not, until a certain number of predictions in a row are observed, i.e. an n-tuple of equal predictions. It can be assumed that with increasing tuple size, i.e. double, triple, quadruple, quintuple, etc., the certainty of the classifier and thus the accuracy of the prediction rises. If such a rule is applied, the performance of the model is increased as depicted in Fig. 6.5.

A number of observations can be derived: first, with an increasing dialog duration and delay of the actual decision, the accuracy of the model rises for all tuples. This can be explained by the growing amount of data that is gathered with the growing progress of the dialog. Second, the accuracy is up to 20 % higher when a quadruple of equal predictions is used as a decision instead of when relying on a single prediction. A tradeoff has to be accepted when using n-tuples as decision rule: not all calls exhibit for example quadruples, i.e. no decision will be reached for these calls. As depicted in Fig. 6.6, in 11.61 % of all calls no decision will apply, when deciding after observing a quadruple. It seems that the classifier often changes the expected outcome and therefore the course of the dialog is marked by great uncertainty. In an SDS employing such a model the decision has to be taken, whether to escalate such calls or whether to take the higher risk that the dialog will not end successfully. This comparatively simple approach lacks the flexibility of being able to control the decision point, since the decision is only triggered by the course of the predictions.

Mean Confidence Deviation Metric

We have shown that the n-tuple approach introduced in the previous section already leads to a large gain of prediction accuracy. We followed up on this idea and tried to view the problem from a different perspective. Instead of coming to a decision when a certain amount of the same consecutive predictions are observed, it might be a more elaborated approach to keep track of the *frequency of changes* of the prediction sequences. In other words: The more fluctuating these sequences are, the less likely are these predictions correct, since there seems to be a greater degree of uncertainty in predicting the correct outcome of this particular dialog. The confidence of the classifier at an exchange E can mathematically be described by summing up these differences divided by the number of exchanges E that have already been processed, also called Mean Confidence Deviation MCD1:

$$\mathrm{MCD1}(E) = \frac{1}{(E-o)+1} \sum_{e=o}^{E} \sum_{\omega=0}^{\Omega} |x_{e,\omega} - x_{e-1,\omega}|, \tag{6.15}$$

where o is an offset that we require since the first three exchanges in our employed dataset contain greetings and introduction prompts that do not contain any predictive power. We thus choose $o = 4$ for the employed dialog system which is the first exchange in each call that differs from the other calls. ω are the possible outcome classes $\in \{0, 1, 2\}$, i.e. SUCCESS, PARTIALSUCCESS, and FAILURE. $x_{e,\omega}$ is the class-wise confidence score (i.e. belonging to class c) at exchange e. It can be seen that in the example dialog in Fig. 6.4 at exchange four the three sequences split up. This happens with every call, since it is the exchange, where the first real prediction is possible. Because of this split MCD1 has already the value 0.66 at this point in the depicted call (cf. Fig. 6.12). If, from this exchange on, the prediction sequences remain steady, MCD1 quickly converges to 0, and every time a confidence change occurs MCD1 diverges from 0. With this characteristic a decision point can easily be introduced by defining a threshold θ. The decision point is then determined, once the condition $\mathrm{MCD1}(E) < \theta$ is fulfilled. However, the metric has a decisive disadvantage as can be easily comprehended when looking at Fig. 6.12: early uncertainty will keep the MCD1 score high. It is thus difficult to fall below θ in later exchanges. The relevance of early uncertainties has to be reduced. We thus further introduced two parameters α and ρ and modified the equation as followed:

$$\mathrm{MCD2}_{\alpha,\rho}(E) = \frac{1}{((E-o)+1)^{\alpha}} \sum_{e=o}^{E} \sum_{\omega=0}^{\Omega} |x_{e,\omega} - x_{e-1,\omega}| \cdot \frac{1}{((E+1)-e)^{\rho}} \tag{6.16}$$

The new variables are:

- An acceleration factor $\alpha \in \mathbb{R}\{0.5\ldots1.5\}$ increases the denominator with an increasing call length. Choosing a larger value here forces an earlier decision.
- A reduction factor $\rho \in \mathbb{R}\{0.5\ldots1.0\}$ decreases the relevance of changes of the prediction sequences, which happened long before the current exchange.

6.5 Evaluation

The linear and window models are assessed using SC-BAC, SC-VAC, LG-FIELD, and LG-LAB. The two automated troubleshooters SC-BAC and SC-VAC respectively contain automatic outcome classes. They have been automatically marked by the LevelOne system after the interaction. They are

- DEFLECTED for calls where the troubleshooting was successful,
- PARTIALLYDEFLECTED for calls where the technical problem of the user was at least identified, but the system failed to complete the troubleshooting, and
- ESCALATED for calls that had to be transferred to a live agent since the system failed plus sudden hang-ups.

Data from the Let's Go system LG-FIELD and LG-LAB has been manually annotated with one of the three classes

- COMPLETED for calls where the system offered the correct schedule,
- FAILEDDUETOSYSTEMBEHAVIOR for calls where the system offered the wrong schedule,
- FOUNDOUTTHATTHEREISNOSOLUTION for calls where no bus schedule could be found meeting the user's query, and
- NOTCOMPLETED for calls ending with sudden hang-ups.

All dialogs of the employed corpora have been assigned to one of the outcome classes $\Omega = \{\text{SUCCESS}, \text{FAIL}\}$ depending on the respective outcome. Thereby, for the troubleshooters the classes PARTIALLYDEFLECTED and DEFLECTED have been mapped to SUCCESS, and ESCALATED to FAIL. In the Let's Go systems, COMPLETED has been mapped to SUCCESS and the remaining three classes to the class FAIL.

The employed input variables to predict the outcome class have been previously been defined. For the linear model, the size of the feature vector amounts to $25 \cdot \gamma$, where γ is the classification point. The window model uses a feature vector of static size not depending on the time of classification and has a fixed size of $25 \cdot w + 19 + 11$, where w is the size of the window, here $w = 3$.

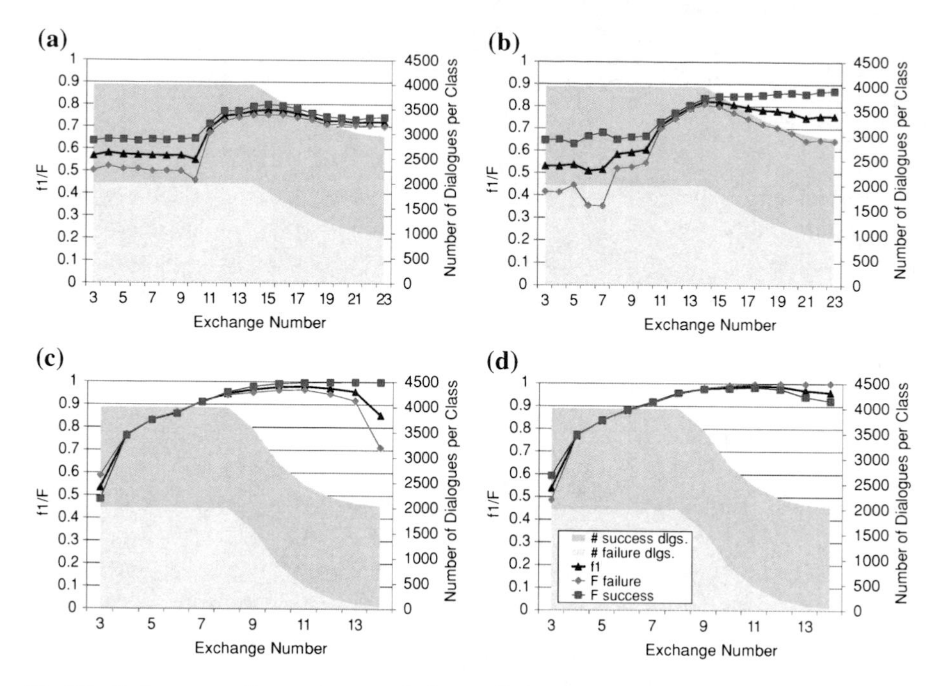

Fig. 6.7 Performance of task success prediction applied on the automated troubleshooters SC-BAC and SC-VAC. *Left side* linear modeling (**a** SC-BAC: linear modeling, **c** SC-VAC: linear modeling), *right side* window modeling (**b** SC-BAC: window modeling, **d** SC-VAC: window modeling)

Due to the limited size of LG-FIELD and LG-LAB all data for training and testing has been used. From SC-BAC and SC-VAC we respectively chose a subset of 4,000 dialogs with equal class distribution. We limit the classification of task success on exchanges 3–21, as for earlier classification points only insufficient data is available, and for later classification points a large proportion of unsuccessful dialogs are already finished. In SC-VAC we confine ourselves on the exchanges 3–14 as the largest proportion of unsuccessful calls is finished after approximately 16 exchanges. It should be noted that the *last two exchanges* of each dialog are *not included* in the feature vector. This ensures that no obvious information about the outcome enters classification and that a dialog manager still has a time-slot of two exchanges for reacting on an estimated task failure.

The classifier employed is an SVM with sequential minimal optimization (SMO) (Platt 1999) as it has shown superior results for this task compared to other classifiers, such as RIPPER, ANN, and kNN. Evaluation has been performed using 10-fold cross validation with stratified sampling for all data sets. The results are depicted in Figs. 6.7 and 6.8.

Next, we assessed the influence of the extended SRI-based feature set as described in Sect. 6.3 using the SC-BAC data set. Therefore, the current exchange that is subject to classification was modeled using the feature vector as described in Eq. 6.14. In

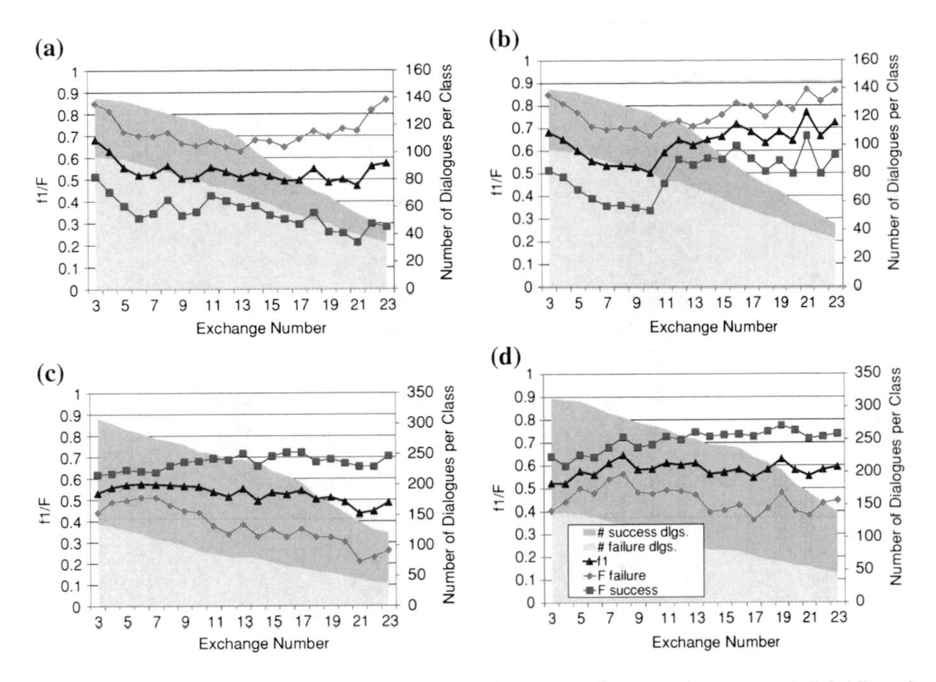

Fig. 6.8 Performance of task success prediction applied on the Let's Go corpora LG-LAB and LG-FIELD. *Left side* linear modeling (**a** LG-LAB: linear modeling, **c** LG-FIELD linear modeling), *right side* window modeling (**b** LG-LAB: window modeling, **d** LG-FIELD: window modeling)

order to prevent invalid information from entering the feature vector, the SRI statistic for the trigrams (see Eq. 6.12) has been determined on the remaining 100,800 dialogs, i.e. without the 4,000 dialogs that are subject to the actual evaluation. Results are shown in Fig. 6.10.

6.5.1 Overall Performance

As becomes visible in the Figs. 6.7 and 6.8, the performance of task success prediction heavily depends on the data set employed. Task success prediction in SC-BAC is close to random in the beginning of each interaction. Obviously it seems not predictable in an early phase, in which direction a dialog is developing. This suddenly changes at approx. exchange number nine, where the performance significantly increases. This performance leap can be attributed to the tree-structure of the dialog flow, as branching at specific points in the tree implicitly decides over task success. This effect appears stronger in SC-VAC, where the performance considerably and constantly increases with the progressing dialog. At exchange number nine, an almost perfect *f1* score for prediction is reached. For the Let's Go systems, the charts (Fig. 6.8) show more diffuse estimations. While task success seems to a certain extent predictable

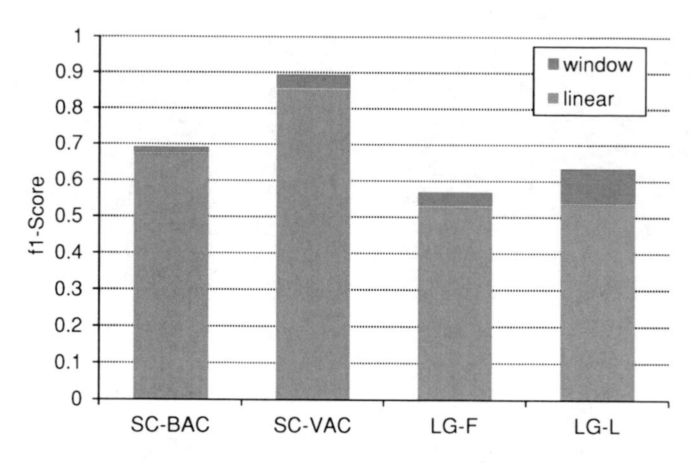

Fig. 6.9 Average performance of task success prediction for both, linear and window model

in the laboratory setting of LG-LAB with an average $f1 = 0.63$ and a maximum $f1 = 0.77$ at exchange 21, it virtually fails for the field data LG-FIELD, where the average *f1* score for all exchanges amounts to 0.57. This can mainly be attributed to sudden hang-ups that are not predictable, and which in contrast do not occur in the lab data. A second explanation of the comparatively poor performance involves that the dialog flow of the Let's Go system resembles less a tree structure as the system handles merely a single task, namely offering bus schedules.

6.5.2 Linear Versus Window Modeling

The advantage of a compact representation of the interaction becomes visible when comparing the linear with the window model. In general, the window model outperforms the linear model in all four data sets. This holds true particularly for predicting at later phases in the dialog, where the large dimension of the feature space leads to a degradation of the linear model. A direct comparison of both models is depicted in Fig. 6.9. The window model shows an absolute performance gain of 0.02–0.09 *f1* when considering the average performance for task success prediction.

6.5.3 Class-Specific Performance

Differences can be further observed when comparing the class-wise *F*-measures. For SC-BAC, SC-VAC, and LG-FIELD, task SUCCESS predicts better than task FAILURE. Clearly the opposite is the case for LG-LAB, where FAILURE predicts better. The signs for failure are more clearly visible than in the three real-life data sets. This involves

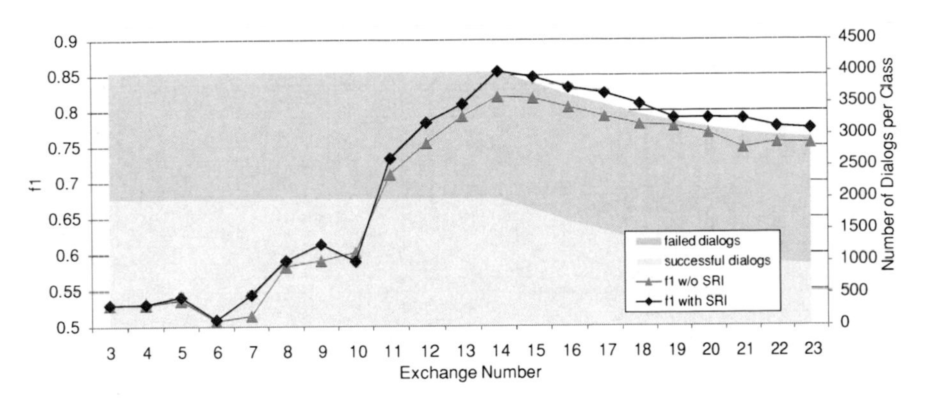

Fig. 6.10 Enhanced window model using SRI features

for example a generally lower rate of average confidence scores for FAILURE dialogs and a higher number of unrecognized user inputs.

6.5.4 SRI-Related Features

As shown in Fig. 6.10, the SRI derived from the trigram statistic into the model, the performance increases by $f1 = 0.018$ in average, which is a relative improvement of 2.6 %. While virtually no performance gain is visible in early exchanges, a considerable increase may be noticed after exchange seven. This effect lies in the nature of the model as the SRI of the trigram at later exchanges better represents the likelihoods of the respective dialog outcome classes.

6.5.5 Model Uncertainty

For assessing the uncertainty formulas MCD1 and MCD2, we numerically searched for the most effective combination of the input variables and iterated over more than 1,200 different combinations of α, ρ, and θ. The complexity is increased since a maximum of the three output variables *accuracy*, *average decision point*, and *percentage of calls* covered by the metric represents a problem. It should be noted that the decision point is not static for all calls, e.g. at exchange number 10. It is dynamically determined once MCD2 falls below θ. We thus speak of the *average* decision point. This dependency is depicted in Fig. 6.11. If we would aim for the highest possible *accuracy*, the logic consequence is that the decision point has to be set as late as possible and/or the decision has to be only taken when the certainty is very high. It is easy to comprehend that a classifier predicting task success only makes sense when it gets a chance at comparatively early stages in a dialog when

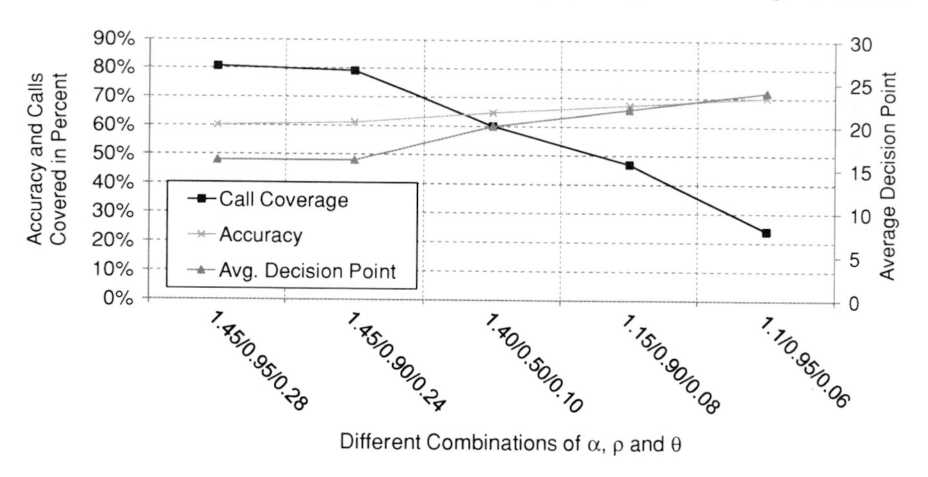

Fig. 6.11 Dependency of accuracy, average decision point, and percentage of covered calls for selected example values of acceleration variable α, reduction variable ρ, and threshold θ, ordered by accuracy

still covering a large amount of calls. For SC-BAC, the optimum combination is $\theta = 0.24$, $\alpha = 1.45$, and $\rho = 0.9$. Obviously this decision is debatable and depends on the system requirements and has to be adapted to each specific system. For this configuration, the average decision point is 16 covering 79.14 % of all calls with an accuracy of 61.17 %. This approach outperforms the n-tuple approach, e.g. a quadruple decision rule would yield 58.5 % with an earliest decision point of 13 and an average decision point of 16 and covers 80.0 %.

6.6 Summary and Discussion

Increasingly, SDS are deployed to resolve complex tasks jointly with the user that go beyond mere information retrieval. Nowadays these dialogs may sometimes last up to 20 min (see Fig. 3.1a, b in Sect. 3.1), such as in automated technical support troubleshooters that help users to resolve technical problems. The complexity of the systems and the limitations in speech technology both represent a continuous threat for a successful dialog outcome. In those circumstances, future SDS must be enabled to detect dialogs, which run the risk of task failure. With that knowledge, users may be brought to task success, for example by help of a live agent or an adapted dialog strategy.

As there exists a complex dependency of the interaction and the probability of task success this problem can hardly be solved by means of static rules[4] and requires

[4] Common heuristic indicators for determining task success is e.g. the occurrence of several subsequent mistaken user utterances, e.g. "IF #ASRREJECTIONS>=3 THEN task_success=false".

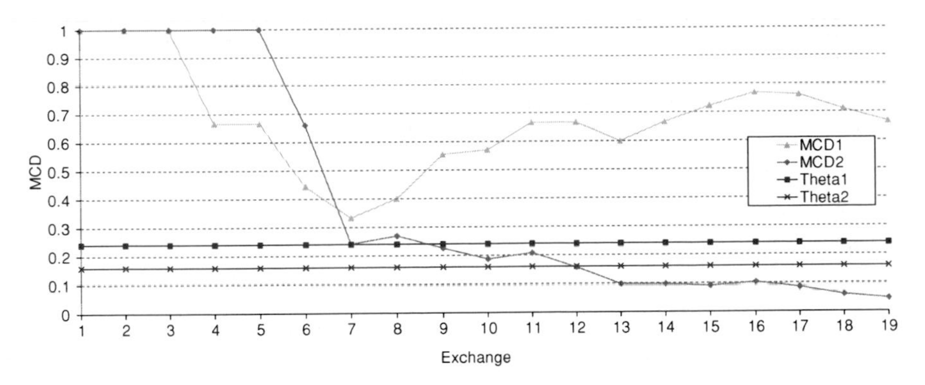

Fig. 6.12 Both metrics applied on the sample call. The threshold $\theta 1$ applies for the MCD1 metric, $\theta 2$ applies for MCD2. Applying MCD2 would trigger a decision at exchange 12, applying MCD1 would not trigger a decision since the call is considered as unsure according to this metric. Both thresholds have been determined using 629 test calls

a statistical solution. In this chapter, we have thus presented statistical models for estimating the dialog outcome in SDS. For example statistical approaches may take implicitly into account input variables that are less obviously linked to task success, such as the current dialog step and the barge-in behavior of the user. The strength of the presented statistical models is furthermore straightforward portability toward other system versions and domains. This has been successfully demonstrated by applying the models on four different SDS.

The proposed solutions follow two different strategies for estimating task success. Both use a pattern-matching approach and rely on large amount of training data. The first *linear* model takes into account the entire previous interaction, which is represented as a feature vector with interaction parameters of each single previous exchange. Every single detail of the previous interaction is hereby used. This procedure brings along an increasing complexity of the feature vector as with incrementing dialog history the dimension grows linearly. Large amounts of training data are required to avoid the curse of dimensionality and to adequately model task success. A solution to this problem could be found by introducing the *window* model that includes the most recent interaction in higher resolution and dialog steps further in the past with window and dialog-wide parameters that summarize the previous interaction. It could be shown that the model performed better particularly when estimating task success in later exchanges of a progressing dialog. By using self-referential information features of the dialog manager's activity we demonstrated that the tree structure of a dialog system may be exploited to improve the model's performance.

Applying task success models on a large number of dialogs and analyzing the model's estimations in the Witchcraft Workbench it could be observed that predicting task success comes with a high uncertainty, as particularly in early phases, a dialog outcome may hardly be correctly estimated. We have proposed the n-tuples and

MCD-metrics to deal with that specific problem. The latter ensures that the model is forced to a decision the further the dialog progresses.

A number of influencing factors can be identified when assessing the validity of the models.

Number of Samples

The performance degrades considerably when an insufficient amount of training material is used. This particularly holds true for the linear model applied on LG-LAB, see Fig. 6.8a. From exchange 13 on, the amount of remaining calls belonging to the class FAIL decreases strongly as many calls belonging to that class are of shorter duration than the calls belonging to the class SUCCESS. The Hughes effect can be avoided when applying the window model, see for example Fig. 6.8b, as the window model is based on feature vectors of lower dimensionality.

Class Label

The definition of task success is strongly domain-specific and can not be generalized for all SDS. For the troubleshooting domain a task is considered as successfully solved when the call has been completely automated, i.e. the user reached task success without external help of an agent. While it seems that dialogs can be clearly separated into successful and unsuccessful ones by this definition, this is far from being the case. Some users may solve a problem, i.e. reach task success, without giving feedback to the troubleshooter. For example users who have lost their Internet connection are led to a solution step by step. Many users who have regained their connectivity hang-up without getting back to the system. For such calls, the task success label may be invalid, which leads to noisy models. In the Let's Go bus schedule domain similar issues may arise as multiple queries in a row are possible. While the first query might meet the user's expectations, the second query might not.

Modeling

The high dimensionality of the feature vector does not necessarily lead to performance loss for linear models if sufficient training data is available. However, training models with larger data amounts increases considerably the duration of the training phase. The performance loss for later exchanges can further be compensated by applying feature selection. We noticed that applying an IGR ranking scheme with feature selection improved the performance of the linear models and reached higher performance scores in later exchanges. Nevertheless, the complexity is drastically risen as the feature selection process has to be conducted on every single decision point. The proposed window model appears to be the better choice as both complexity and performance are superior. There may be other modeling approaches that may better meet the sequential nature of the classification task. Instead of modeling each interaction in a static feature vector, the interaction may be modeled likewise as sequence of observation vectors that may be classified using a single Hidden Markov Model for all decision points. Details on the presented modeling approaches are described in Schmitt et al. (2010f).

Model Uncertainty

It could be shown that predicting task success in an ongoing spoken dialog interaction brings along uncertainty that has to be dealt with. Certainty, and thus accuracy of the model, can be increased by relying on a series of homonymic hypotheses instead of a single prediction alone. The decision point is thus determined by the first occurrence of such n-tuples with subsequent identical predictions. The employed SC-BAC dataset exhibited an average increase in accuracy of 6.1 % for doubles, 10.9 % for triples, and 11.8 % for quadruples. Quid pro quo: in return for the rising accuracy, we cannot come to a decision in all dialogs. Respectively, 1.44, 5.0, and 11.61 % of the dialogs exhibit no doubles, triples, or quadruples, respectively at the first possible decision point. However, it should be reminded that without a statistical model, even in 100 % of the calls there would be no decision to which outcome a call will lead. An even higher accuracy can be yielded by introducing the Mean Confidence Deviation metric that keeps track of the classifier's confidences. The metric considers a decision as uncertain, as long as the confidences for the single classes rapidly change during a dialog. Depending on the acceleration variable α a decision is forced earlier or later in the course of a dialog. The reduction variable ρ limits the influence of early uncertainties. Both variables influence the earliness and the accuracy of the decision as well as the number of dialogs where a prediction is possible. Details on this book contribution may be found in Schmitt et al. (2011c).

Chapter 7
Conclusion and Future Directions

Spoken Dialog Systems (SDS) are being deployed with ever-increasing frequency while rapidly gaining complexity. Today's systems mostly follow a static dialog flow and offer static prompts to all users, which results in perceptible unnatural and sometimes even unsuccessful interactions. "Adaptivity" is therefore a key concept in the next years that will dominate spoken dialog research (see e.g. Langley 1999; Langley et al. 1997; Riccardi and Gorin 2000; Müeller and Wasinger 2002; Litman and Pan 2002; Pittermann et al. 2009; Heinroth et al. 2010a; Lemon and Pietquin 2011; Bezold and Minker 2011; Rieser and Lemon 2011). Future SDS will be enabled to adapt their behavior to the user's needs based on user models and interaction-related knowledge.

We consider adaptivity a two-step process as proposed with the Adaptivity Wheels, cf. Fig. 1.4 in Sect. 1.2. The term "Adaptivity" may be split hereby into two central phases. In the *detection phase*, we determine dynamic and static user properties as well as properties adhering to the ongoing interaction. For realizing the detection, recognizers endowed with statistical models need to be implemented that allow the estimation of specific target variables based on input variables. The input variables themselves are derived from observations made during the interaction. In the *action phase*, the estimated information is required, where the actual adaptivity takes place. Adaptivity may for example be reached through choosing speaker-specific acoustic models, an adapted language style for system prompts, altered confirmation strategies or simply seeking human assistance, e.g., by transferring the user to a live agent in an SDS call-center application.

The book at hand has solved a number of issues related to adaptivity with particular focus on the detection phase. Special emphasis has been put on the **statistical modeling for the detection of critical dialog situations**. By "critical" we mean dialogs that show signs of poor communication and run the risk of ending with an unsatisfactory result, which may be task failure or unsatisfied users. In this respect, most current SDS are overwhelmingly inflexible and constrained. The vision and the standard for the next generation of SDS should be one that is more dynamic by detecting critical interactions and ultimately solving such problems by changing the

A. Schmitt and W. Minker, *Towards Adaptive Spoken Dialog Systems*,
DOI: 10.1007/978-1-4614-4593-7_7,

dialog strategy. This may lead to an increased task success rate and a higher user satisfaction.

The presented approaches have been of statistical nature. Hereby, concepts and classes (e.g., satisfied vs. dissatisfied, angry vs. non-angry) are automatically learned from large annotated training corpora and memorized in the form of model parameters. These parameters are then used by the model to determine the most likely concept or class (e.g., emotional states, dialog outcome, user satisfaction) that fits best the new observations. In contrast to static hand-crafted rules, the statistical approach allows for an easy portability to other system versions and domains. The book at hand has successfully demonstrated this using four different corpora from various origins and domains.

7.1 Overall Summary and Synthesis of the Results

The examples in the introduction in Tables 1.1 and 1.2 have demonstrated that communication problems may occur in SDS leading to unsuccessful dialog outcomes and unsatisfied users. From then on, we have analyzed different aspects adhering to troublesome communication and developed novel techniques for their detection. The proposed approaches may be unified under the roof of an Online Monitoring Module (OMM) that communicates the observations made to the dialog manager in form of a *knowledge vector*. Based on the knowledge, certain actions may be triggered. Thereby the presented knowledge vector is not only restricted to information related to critical dialog situations, but may also entail estimated properties of the user, i.e., an estimated user model. Applying the proposed approaches to the real-life example from Table 1.1, a conversation between system and user may then go off as depicted in Table 7.1.

The novel OMM is able to monitor the interaction between the system and the user. Based on input variables derived from the interaction consisting of acoustic, prosodic, linguistic, and interaction-related parameters, the module is able to estimate static and dynamic user properties and characteristics about the interaction. The OMM itself consists of several submodules entailing recognizers that are based on statistical models and that have been partly implemented in this book. In this example, the module estimates the outcome of the dialog (SUCC/FAIL), the Interaction Quality as well as the gender of the speaker at each single dialog step and its emotional state and age.[1] Using the knowledge generated by the OMM, the dialog manager may trigger adaptivity steps to render the interaction more dynamic, user-friendly, and robust.

[1] Speaker age estimations are particularly error-prone (Müeller and Burkhardt 2007; Metze et al. 2007). It may thus be advisable to postpone a decision about the speaker age until sufficient amount of speech data are available allowing for more reliable age estimation. Due to uncertainty caused by the short sample length available to the age model in the beginning of the dialog, the age entry of the first exchange is thus marked with a "?".

Table 7.1 Possible adaptivity scenario using the knowledge vector generated by an OMM

Interaction	Knowledge vector				
	(TS	IQ	EMO	Gender	Age)
S: Welcome to the CMU Let's Go bus information system.					
To get help at any time, just say Help or press zero.					
What can I do for you?					
U: 54C schedule *[54C]*	(SUCC	5	Neutral	Male	?)
S: The 54C . Did I get that right?					
U: Yes. *[yes]*	(SUCC	5	Neutral	**Male**	<20)
S: Alright. Where do you want to leave from, **my young friend**?					
U: West North Avenue *[/noise/ /mumble/]*	(SUCC	5	Neutral	Male	<20)
S: Leaving from EIGHTH AVENUE . Did I get that right?					
U: No, West North Avenue. *[eighth_avenue]*	(FAIL	4	Neutral	Male	<20)
S: Leaving from EIGHTH AVENUE . Did I get that right?					
U: No! *[north_avenue]*	(**FAIL**	**3**	**Angry**	Male	<20)
S: Sorry, I am annoying you buddy, but I can hardly grasp what you mean. Please tell me a neighborhood nearby.					
U: Eastfield.	(FAIL	3	SlightAng	Male	<20)
S: Leaving for WESTNORTH AVENUE . Did I get that right?					
U: yes	(SUCC	4	Neutral	Male	<20)
S: Okay. Where are you going?					
U: To Whigelow	(SUCC	4	Neutral	Male	<20)
S: ...					

The knowledge vector in this scenario contains task success prediction (TS), which is either SUCC for success and FAIL for task failure. Further, it contains the estimated interaction quality (IQ) on a score from 1 to 5, the estimated emotional state of the user (EMO) and gender and age classes. The latter two submodules have not been discussed in this book. Details to age and gender detection may be found in Schmitt et al. (2009a, 2010d)

7.1.1 Corpora Creation, Basic Modeling, and Tool Development

The suggested online monitoring techniques rely on statistical models that are trained with large amounts of data that required parameterization and annotation. Thus, a number of basic contributions had to be implemented beforehand for being able to develop statistical models.

An interaction parameter set One of the first contributions of this book was the *implementation of a parameter set* that allows modeling an interaction on the exchange level. The resulting set served as input variables for many of the presented approaches. This set complements the parameters proposed by the ITU (ITU 2005) with exchange-level parameters and allows a completely automatic quantification of the interaction.

New dialog corpora In the following, raw data from three different systems and domains have been parameterized and annotated, resulting in *four corpora*. They have been used for assessing the proposed online monitoring models and for analyzing dependencies, e.g., between emotions and user satisfaction. Some of the corpora contain up to 100,000 real-life dialogs. During this work, the largest English-language corpus for real-life anger has evolved containing more than 22,000 utterances. The publication related to this book contribution is Schmitt et al. (2010g).

Software framework For handling and annotating the large amounts of data, a *platform-independent, multi-user software framework* has been developed and has been made publically available as open-source project. The presented Witchcraft Workbench allows first of all an in-depth analysis of logged dialogs by making dialog scripts browsable showing relevant parameters jointly with the dialog flow while playing back the interaction. Several dialog corpora may be managed at the same time; sorting and grouping dialogs according to specified criteria is possible. Furthermore, Witchcraft allows an assessment of discriminative and regressive classifiers (the online monitoring modules) that are intended to provide new knowledge to the SDS. Jointly with the dialog flow, it may now be assessed what the respective modules (e.g., an emotion recognizer) predict at which point in time. It may be further analyzed, which knowledge an SDS has gained through the deployment of a statistical model. Thereby, it may for example be visualized, at what point in time the model seems to have correctly predicted the gender of the user. The concept of the Witchcraft Workbench is to provide an extendable basis for research on adaptivity. The architecture of Witchcraft thus allows the implementation of third-party plug-ins that may contribute to the functionality of the entire framework. Related publications are Schmitt et al. (2010a, b).

7.1.2 Emotion Recognition

Users react emotionally when interacting with computer systems. This holds true in particular when users face problems with an SDS or express their general dissatisfaction about a speech-based automated service hotline. At the beginning of this book we have thus addressed the most obvious symptom for poor communication, namely the occurrence of negative emotions. It turned out that users interacting with today's field SDS usually do not show emotions except negative ones. Other emotions such as happiness, disgust, fear, etc., could not be observed and may thus be considered irrelevant for the further detection phase. It should be noted that this statement applies for today's SDS. By nature, the "machine" we are nowadays interacting with is an anonymous interlocutor that we mostly connect to by phone. With the coming of new applications such as personal software assistants this may rapidly change. For example, Siri from Apple and related applications that may build a "personal" relationship with the user may elicit other emotions than negative ones.

As for today's telephone-based SDS, strong differences in the distribution of anger in SDS could be noticed when analyzing different domains. While only very few users reacted angrily during an interaction with the automated troubleshooter (SC-BAC), a larger proportion of users showed anger in the bus information system (LG-FIELD, LG-LAB). Related work offers little information on emotion distributions. Liscombe et al. (2005) reports 73.3 % non-angry speech in the HMIHY 0300 corpus, which in turn means that 26.6 % contain negative speech. In contrast, Burkhardt et al. (2008) report 7 % anger in a commercial SDS. It is evident that the definition of "anger" is subject to strong variations among the related work and ranges from clear, hot anger to a mixture of slight to strong negative emotions.

Apart from the varying definitions of what is considered anger, the reasons for the varying distributions among the systems may be of diverse nature, such as system design, choice of the computer voice, context of use etc. A very likely cause, however, is the size of the grammar affecting some input slots. For example, in the Let's Go system users may enter a large number of bus stop names and neighborhoods which again raises the probability of confusions. The automated troubleshooters have very restricted grammars, which limits the risk of confusions. The Let's Go system further has to deal with a generally higher background noise level as users of Let's Go call the system while being on the move. Both, the larger grammars and the higher background noise level are reasons for frequent ASR rejections, which again causes displeased users. The absence of background noise may be one reason why users from the LG-LAB set showed less anger than their counterparts in the field. It seems natural that the usage of emotion detection in SDS is only justified when a certain amount of emotions is present. Yet this is not always the case in many data sets, where the emotion distribution is extremely skew toward non-anger classes. Given that emotion detection remains error-prone (expressed through the low precision scores), the risk of misclassifying an emotionless utterance as an emotionally charged one is high. Particularly, in interactive voice response (IVR) systems, user prompts are frequently very short and rarely last longer than two seconds, which further complicates a reliable detection of the user's emotional state.

Before deploying emotion recognition in an SDS, it should thus be first estimated if a possible adaption strategy following the detected emotion may lead to higher task completion rates and/or higher user satisfaction. Second, the emotion distribution should be analyzed before implementing a domain-specific emotion recognizer. For

[htb]

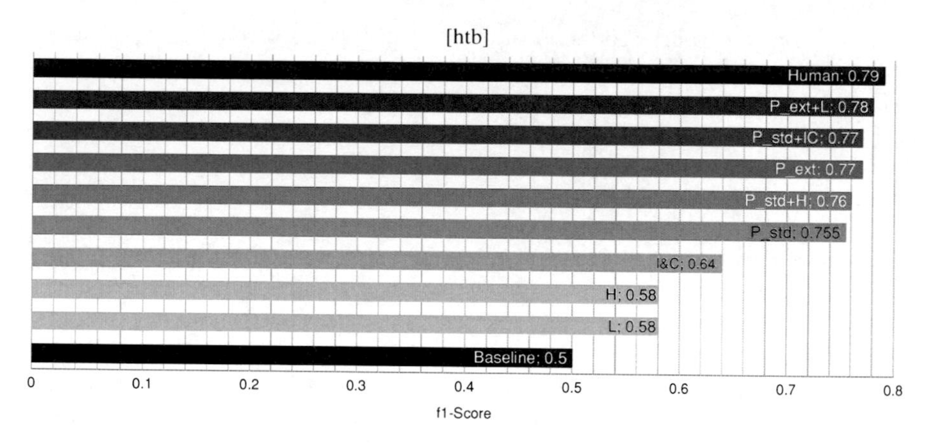

Fig. 7.1 Comparison of all emotion recognition techniques applied on SC-BAC

example, if users rarely show anger in the respective scenarios then the use of anger recognition would be without effect.

Previous studies dealing with emotion recognition are often based on limited acoustic models and rule-based linguistic models and are frequently evaluated with unnatural, artificial, and acted data sets. The aim of such studies is the development of models suitable for a general recognition of emotions in speech, i.e., they are not necessarily intended for SDS. Thus, they neglect information exceeding the frontier of a single spoken utterance. This information, however, may contribute to emotion recognition in SDS. The book at hand explored new features and knowledge sources on real-life data that have not been analyzed in previous studies. Apart from mere paralinguistic and linguistic information, further knowledge has been derived from the interaction. The performance of all presented models applied on SC-BAC is depicted in Fig. 7.1. The single modules are summarized and discussed in the following.

Acoustic Modeling

The most obvious characteristics of human emotions are of acoustic nature. In this work, we have presented two acoustic models; one model that resembles state-of-the- art approaches (P_{std}), with 52 acoustic and prosodic features, and an extended model (P_{ext}) that takes into account further variations in speech, resulting in a 1,477-dimensional feature space. When feature selection is applied, this latter "brute-force method" achieves recognition rates ranging from $f1 = 0.68$ (AIBO) to 0.77 (SC-BAC), which is a relative performance increase compared to P_{std} of 1.3 % (SC-BAC) to 5.6 % (MOB). It should be noted that the computational effort for extracting the larger feature vector for P_{ext} is comparatively high. Hence, it seems sensible to use the basic acoustic model P_{std} in deployment scenarios and to add less computationally complex interaction- and context-related features, which compensate the less detailed acoustic model while achieving a similar performance as P_{ext}. Both acoustic models are based on feature vectors of static size not depending on the length of the utterance. While this may be sufficient for short utterances of 1–3 s common in IVR systems, the approach might achieve suboptimal results as it might not capture all acoustic properties of longer utterances. Chunking longer utterances into shorter ones could then be a possible remedy. Experiments using dynamic approaches such as Hidden Markov Models as described in Pittermann and Schmitt (2008) did achieve suboptimal results on the described data sets.

For classifying the static feature vectors best results could be achieved with SVMs, which outperformed other base classifiers such as rule learner, kNN, and ANN as well as the employed ensemble methods bagging, boosting, and voting. The results confirm the findings from Schuller (2006).

Linguistic Modeling

For linguistic emotion recognition, we explored statistical approaches and assessed vector-space models in particular, such as BOW, TF, and TF·IDF that are originally intended for document classification. With the emotional salience, we furthermore assessed an entropy-based model for suitability on real-life data. It could be shown that linguistic emotion recognition does not yield satisfying results for most current real-life SDS as common utterances from today's systems are of very short command-like nature and do not entail enough linguistic expressiveness. It should be noted that linguistic information indeed may considerably increase overall recognition performance, for example in human–human spoken dialogs or more complex user utterances of longer duration in future SDS. Comparing hand transcripts with (erroneous) ASR transcripts we could show that the linguistic models exhibit a high robustness toward ASR errors. Nevertheless, it is advisable to train the linguistic model with hand transcripts instead of erroneous ASR transcripts, as the latter model achieved a $f1 = 0.02$ lower performance. The best among the tested approaches are the entropy-based EmoSal models, which, however, stay far behind the acoustic models achieving merely $f1 = 0.58$ (SC-BAC) up to $f1 = 0.64$ (MOB). Although our results are based on three real-life corpora allowing for a more general statement on the capability of linguistic information for anger detection, more positive results for linguistic models have been reported by Lee et al. (2005). They employed EmoSal on comparable IVR data from the AT&T HMIHY system. Fusing acoustic with linguistic information they were able to reduce the classification error by 7–9 % absolute over using only acoustic models. Unfortunately, no class-wise F measures are given in their study. The provided classification errors may be heavily influenced by the skew distribution and the fact that speaker-dependent models have been used.

Interaction- and Context-Related Modeling

Further attention has been paid to the interplay of interaction and emotion. While for example poor speech recognition performance may elicit negative emotions, this implication turned out to be helpful for recognizing emotions, as knowledge about certain system events themselves may indicate their occurrence. A correlation analysis has shown positive dependency with the degree of anger and certain interaction parameters. Recurring reprompts thereby seem to particularly annoy users. Other observations related to anger may further be the duration of the user utterance, which tends to be of longer duration when anger is expressed. The presented model exploiting these phenomenona is based on interaction parameters representing the previous interaction. For both assessed data sets, this model, based entirely on non-acoustic and non-linguistic information, achieves $f1 = 0.64$ and thereby outperforms all linguistic models. It could be further shown that the validity of the model holds partially true when it is applied on data from completely other domains. Even if an implementation of interaction- and context-related model seems to be too complex for some domains, it is advisable to take into account some ASR performance scores as additional features complementing the acoustic and prosodic classifiers. The suggested features do not require manual annotation, as e.g. in Lee et al. (2005).

History-Related Modeling

We could show that negative emotions do not occur in isolation in the course of spoken dialogs. The observation of a negative emotion implies high likelihoods for observing another negative emotion in the next exchange. Taking this into account, a statistical model may contribute to predict the actual emotional state at the current exchange using prior information about the user's emotional state. The presented Hidden Markov Model allows predicting the negative emotional state of the user solely based on information about the previous states. It achieved $f1$ scores of 0.58 (SC-BAC) and 0.74 (MOB). Generally, it can be assumed that the higher the model performance the stronger the implication that a negative emotion follows another negative one. A history-related model may thus particularly be beneficial if this prerequisite is fulfilled. While the proposed HMM is based on the discrete class labels A and N as observations, even better results might be achieved if the observations are acoustic characteristics of an entire utterance, as, e.g., the static feature vector of P_{std}.

Fusion

As the paralinguistic model showed the highest performance among all suggested models, it depicts our baseline. All proposed approaches successfully contribute to a performance increase when applying late or early fusion with the paralinguistic baseline model. Fusing the extended paralinguistic model with the linguistic model achieved a relative performance gain of 1.3 % (SC-BAC) to 2.9 % (AIBO). The presented interaction- and context-related model for recognizing emotions results in a relative improvement of 1.3 % (SC-BAC) up to 10 % (LG-FIELD) over the paralinguistic model. The history-related model contributes up to 8.3 % relative performance gain (MOB).

While emotion recognition may achieve satisfying performance scores on balanced data sets, very skew distributions of the emotion classes may occur in real-life applications. This may result in a large number of false positives. Our answer to this issue was the introduction of a cost-sensitive classifier, which successfully reduced the number of false positives to a large degree while only slightly increasing the number of false negatives. The comparatively poor precision and recall measures for the anger class remain an issue in real-life emotion recognition. Further research will be required that focuses on the specific increase of the anger class performance.

The publications related to this book contribution are Polzehl et al. (2009a, 2010a, b, c, 2011), Schmitt et al. (2009b, 2010c, d, e, g), Schmitt, and Polzhel (2010).

7.1.3 Interaction Quality

Recognizing negative emotions for achieving adaptivity in SDS anticipates that these emotions are signs of dissatisfaction. While it may be valid that users expressing negative emotions are dissatisfied, the inverse conclusion has been proven to be wrong, namely that *all* dissatisfied users express negative emotions. According to the statistics derived from the presented lab study, only a minority show negative emotions when they are not satisfied. In order to be able to identify such dissatisfied users with the aim to adapt the dialog strategy, other techniques than emotion recognition are of need.

Interaction Quality Framework

Our solution was the pattern-based Interaction Quality framework that has been developed in accordance to the PARADISE approach (Walker et al. 2002). While PARADISE is intended to estimate system quality based on a regression analysis from parameters describing completed interactions, we shifted the paradigm to the exchange level, allowing to estimate the quality of the interaction anytime. The previous interaction is modeled in condensed form as an interaction parameter vector serving as input for the model. In the first instance, the target variables have been manual annotations of expert raters providing scores on the perceived Interaction Quality. This expert-based annotation of dialogs hereby roughly follows the idea of Evanini et al. (2008). However, in Evanini et al., experts provided scores for *completed* dialogs, not for single exchanges. An objective view on the interaction quality is assured by creating the median of several expert scores and by introducing a *ruleset* that allows reproducibility. The hypotheses of the user-independent model based on *completely automatic features* achieves strong correlations with the annotated quality scores of IQ_{field} ($\rho = 0.79$) and IQ_{lab} ($\rho = 0.84$).

User Satisfaction Study

In a second instance, a user study with 46 users has been carried out further to analyze subjective user satisfaction with the objective quality annotations, answering the question to which extent such objective IQ scores mirror the subjective impression of the user. It could be shown that subjective user satisfaction shows moderate to strong correlations with objective IQ expert quality scores ($\rho = 0.66$). However, in contrast to IQ, subjective US scores are more difficult to predict as users tend to show strong variations among each other in the perception of deficient interactions. The statistical model's hypothesis using automatic features achieves lower, but still strong correlations with real US scores ($\rho = 0.65$ for US_{lab}). User-dependent models hereby showed better results than user-independent models. However, user-dependent models may only apply for systems where the user is known, which is not the case for most IVR systems. Another major problem related to user satisfaction modeling is its traceability as user satisfaction during an interaction in the field can hardly be measured. The user study further provided information on the correlation between interaction parameters and user satisfaction. Moderate correlations could for example be found for the re-prompt rate,[2] which implies that users get dissatisfied when being prompted several times for the same matter. Weak correlations could furthermore be found with the negative emotional state. Adding hand-annotations about the user's emotional state as feature does slightly improve the performance in all analyzed models.

Apart from estimating the quality of the interaction at arbitrary points in time during an interaction, the model may further serve developers for spotting badly designed dialog steps in the redesign phase. Furthermore, the proposed approach may be used to assess overall system quality and might achieve higher scores than the PARADISE approach as it models the interaction and satisfaction on a more detailed level. The evidence for the latter assumption, however, must still be produced.

The publications related to this book contribution are Schmitt et al. (2011a, b).

7.1.4 Task Success Prediction

Many SDS are intended to automate tasks, such as identifying call reasons in call-router applications, reserve a hotel, and to resolve technical problems.As speech technology still has strong limitations, due to the complexity of speech recognition and semantic interpretation, a large proportion of dialogs ends up being unsuccessful. Detecting that a task is likely to fail may allow an SDS to adapt its strategy in order to

[2] %REPROMPTS

ultimately bring the user to a successful outcome, with or without human assistance. Two different statistical approaches have been developed to model task success. Both cases use training data from previously observed patterns from a large number of dialogs. Based on the statistical models derived from that data, we map a specific pattern representing an interaction of an unknown dialog to one of the outcome classes. Hereby, the patterns representing a dialog up to a certain point may represent every detail of the previous interaction. We called this modeling approach "linear modeling" as the size of the feature vector linearly increases with progressing dialog. The drawback of this approach has been the high dimensionality of the vector. Given insufficient training material in later points the model was dominated by the Hughes effect and at later exchanges, the model performance thus degraded. By introducing the "window model", we confined ourselves on the most recent interaction within a specified window. The interaction from earlier exchanges beyond the window is merely summarized with particular parameters that, for example, give information on the average ASR performance, barge-in rate, etc. This window model showed higher robustness in predicting task success in later exchanges as the dimensionality of the feature vector stays of constant size.

The predictability of task success turned out to be strongly dependent on the employed data set. In average, task success performance ranged between $f1 = 0.57$ (LG-FIELD) and $f1 = 0.89$ (SC-VAC). Obviously, the more the tree representing a dialog structure is branched, such as for the automated troubleshooters, the better the predictability. It should be noted that in most manually analyzed dialogs it could virtually not be foreseen at many points in the dialog whether it would end up successfully or not. This implies that without obvious indicators it is just as difficult for a statistical model to estimate task success. Whether such a model succeeds, thus strongly depends on the respective data. Moreover it is highly unlikely that the presented models generalize, i.e., models trained on one corpus will fail on another. Comparisons to the few related studies targeting on the prediction of task success are not meaningful as data and classification points differ (Kim 2007; Langkilde et al. 1999; Walker et al. 2002; Paek and Horvitz 2004).

It could be further shown that task success prediction comes along with high uncertainties and is subject to strong fluctuations in between adjacent dialog steps. The certainty could be increased by postponing a decision about task success until several identical predictions in a row are observed (n-gram model). A more flexible approach has further been introduced that takes into account the fluctuations of the classifier's class-wise confidence scores. When the fluctuations drop below a predefined threshold, the current decision about the dialog outcome is validated and active. How rapidly a decision is reached can be influenced by an accelerating factor that may be freely chosen.

The publications related to this book contribution are Herm et al. (2008), Schmitt et al. (2008a, 2010f, 2011c).

Finally, other publications resulting from work related to this book in general are Zaykovskiy and Schmitt (2007, 2008b), Pittermann et al. (2008), Schmitt et al. (2008b), Heinroth et al. (2009), Bertrand et al. (2009), Zaykovskiy et al. (2007),

Zaykovskiy and Schmitt (2008a, c), Heinroth et al. (2010a, b), Mowafey et al. (2009), Pittermann and Schmitt(2008), Ultes et al.(2011a) and Zgorzelski et al. (2010).

7.2 Suggestions for Future Work

In this book, we have placed special focus on the detection part of adaptivity in order to allow for online monitoring of SDS. We render SDS "intelligent" by introducing statistical models that aim to predict critical situations in spoken human–computer interaction. In a similar manner as the ASR, which recognizes spoken words, the implemented recognizers provide details to the dialog manager based on observed acoustic, linguistic, and interaction-related parameters, which are tracked during the interaction. Future research should further deal with the implementation and evaluation of novel recognizers and models. The architecture proposed in this book (see Fig. 1.5) may allow an integration of further novel submodules, which may be implemented based on the input features provided in this book. These modules may help to render the interaction more user-friendly. Some suggestions for such models are:

- Discriminating *experts from non-experts* (i.e., novices);
- estimating the user's *willingness to cooperate*;
- recognizing, whether the user is *intoxicated* (as, e.g., in Ultes et al. 2011b; Schuller et al. 2011);
- estimating the *user's educational level* (as e.g. in Zablotskaya et al. 2011a; Zablotskaya et al. 2011b);
- estimating *age and gender* (as e.g. in Schuller et al. 2010);
- recognizing *non-native users* (as e.g. in Goronzy 2002);
- introducing *fatigue recognition*.

By this approach even user models of initially unknown users may be incrementally built up with an increasing duration of the interaction. It should be noted that such recognizers do not necessarily require features originating from spoken dialog. Where possible, video and sensor systems may add further information and support the creation of user models. For example, in automative SDS, fatigue may also be detected by tracking the steering movements or tracking the blink rate of the driver. In kiosk terminals or robotic applications, a user model may additionally be built up using visual sensor data.

We have mentioned that appropriate actions following a detected poor communication may be an adaptation of the dialog strategy or the transfer of the user to a live agent (in call-center applications). Thereby it has not been further specified in the course of this book, *how* this latter adaptation may happen, i.e., the *action* following the *detection* remains undefined (apart from escalating poor calls to call center agents).

Future work will have to explore, how the proposed statistical models integrate into spoken dialog. A statistical approach toward dialog management in this respect has been proposed by Ultes et al. (2011a), which relies on partially observable markov decision processes (POMDP). Furthermore, it needs to be assessed, which strategies may be beneficial for the entire interaction, and more precisely, which strategies ultimately help to raise task success rates and user satisfaction. We have examined some repair strategies in a further study that has not been described in this work. For further reading refer to Zgorzelski et al. (2010). However, more effort is required in the future to examine and define repair strategies. User tests for assessing both the benefit of online monitoring models in combination with adaptivity strategies in SDS will be required.

The pattern-based interaction models for predicting task success and Interaction Quality rely both on static feature modeling, i.e., the interaction-history is represented as single feature vector of static size. In emotion recognition tasks, this static modeling in combination with SVMs has shown superior results over dynamic approaches, which use HMMs. However, on the specific tasks of predicting task success and estimating Interaction Quality, it may yield suboptimal results. In future work it should thus be explored, whether a dynamic model based on HMMs outperforms the presented SVM approaches.

7.3 Outlook

The rapid improvements and developments in mobile and automotive applications will have far-reaching implications on spoken dialog technology. Although the haptic interaction has particularly improved on smartphones compared to traditional consumer mobile phones, the interaction remains cumbersome on the small screens. The fast increase in computing power of smartphones will allow new applications that require novel interaction paradigms. Spoken and multimodal interaction will remain the ultimate goal for interacting with these devices as it is the most natural and immediate way to communicate. The advancement of smartphone technology will also have far-reaching consequences for today's IVR systems that serve customers via telephone. The exclusive speech-based interaction on such telephone-based systems will give way to more sophisticated approaches allowing customers to get in touch with a company or a service via apps for smartphones and tablet PCs,[3] allowing both, visual feedback and multimodal input.

Internet connectivity is bringing applications of completely different nature into our vehicles that go beyond classic board entertainment and navigation. In a few years time, weather apps, news services, social network applications, e-mail, calendar, etc., will belong to the standard services that we expect our vehicles to provide while we are driving. As traditional haptic interfaces depict safety-critical problems, new interaction paradigms are equally required, which can be mainly solved by endowing in-vehicle infotainment systems with a spoken dialog interface. Enquiries, such as "Direct me to the nearest cheap gas station." may then be conveniently solved with

[3] such as "SmartCare Mobile", see www.speechcycle.com.

spoken interaction. The rising complexity of such queries will further demand a real dialog-based interaction between the system and the user going beyond mere command-and-control.

Looking further ahead, we will likewise interact with kiosk systems, humanoid robots, and intelligent homes in a similar way as we do with human counterparts today. In such particular applications, other input modalities than speech are less eligible. For rendering such SDS as natural and intelligent as any human conversation, we will require further advancements in statistical online monitoring techniques allowing us to endow SDS with intelligence and, ultimately, achieve adaptivity.

Appendix A
Interaction Parameters for Exchange Level Modeling

A.1 Interaction Parameters Derived from Automatic Speech Recognition

Table A.1 Interaction parameters derived from the ASR module for employment on the exchange level

Parameter	Description
GRAMMAR	Names of all activated grammars EL: Langkilde et al. (1999) and Walker et al. (2002) DL: Litman et al. (2000)
TRIGGERED GRAMMAR	Name of grammar that returned the ASR parse EL: Langkilde et al. (1999) and Walker et al. (2002) DL: Riccardi and Gorin (2000)
UTTERANCE	ASR parse from user utterance, i.e., automatically transcribed EL: Langkilde et al. (1999), Walker et al. (2002) and Levin and Pieraccini (2006)
ASRRECOGNITIONSTATUS	Status of the ASR when trying to parse the user input $\in \{success, reject, timeout\}$ "success" refers to a—from the ASR's point-of-view—successful recognition of the utterance, i.e., the decoded word string matches an active grammar, "reject" means the ASR could not recognize the utterance and did not find a corresponding word sequence according to the active grammars; "timeout" indicates that the user did not respond within a given time slot EL: Langkilde et al. (1999), Walker et al. (2002) and Levin and Pieraccini (2006)
#ASRSUCCESS	Number of successfully parsed turns up to this exchange e_n $\sum_{i=1}^{n} x \begin{cases} 1 & \text{ASRRECOGNITIONSTATUS} = \text{SUCCESS, R} \\ 0 & \text{otherwise} \end{cases}$ EL: Paek and Horvitz (2004) DL: ITU (2005)

(continued)

A. Schmitt and W. Minker, *Towards Adaptive Spoken Dialog Systems*,
DOI: 10.1007/978-1-4614-4593-7,

Table A.1 (continued)

Parameter	Description
{#}ASRSUCCESS	Number of successfully parsed turns within previous w turns prior to e_n, where w is the size of the window $\sum_{i=n-w}^{n} x \begin{cases} 1 & \text{ASRRECOGNITIONSTATUS} = \text{SUCCESS} \\ 0 & \text{otherwise} \end{cases}$
%ASRSUCCESS	Percentage of "Success" turns in all previous exchanges: $\frac{1}{n}\sum_{i=1}^{n} x \begin{cases} 1 & \text{ASRRECOGNITIONSTATUS} = \text{SUCCESS} \vee \\ \text{ACTIVITYTYPE} \notin \{\text{QUESTION}, \text{CONFIRMATION}\} & \\ 0 & \text{otherwise} \end{cases}$ EL: Paek and Horvitz (2004) DL: Litman et al. (1999), Litman and Pan (2002) and ITU (2005)
#TIME-OUTPROMPTS	Number of time-out events up to this exchange e_n. $\sum_{i=1}^{n} x \begin{cases} 1 & \text{ASRRECOGNITIONSTATUS} = \text{TIME} - \text{OUT}, \\ 0 & \text{otherwise} \end{cases}$ EL: Paek and Horvitz (2004) and Kim (2007) DL: Kamm et al. (1998), Litman and Pan (1999) and ITU (2005)
{#}TIME-OUTPROMPTS	Number of time-out turns within previous w turns prior to e_n, where w is the size of the window $\sum_{i=n-w}^{n} x \begin{cases} 1 & \text{ASRRECOGNITIONSTATUS} = \text{TIME} - \text{OUT}, \\ 0 & \text{otherwise} \end{cases}$
%TIME-OUTPROMPTS	Percentage of "time-out" turns in all previous exchanges: $\frac{1}{n}\sum_{i=1}^{n} x \begin{cases} 1 & \text{ASRRECOGNITIONSTATUS} = \text{TIME-OUT} \\ 0 & \text{otherwise} \end{cases}$ $x = 0 \; \forall$ exchanges in the beginning of the dialog where no ASR is active EL: Paek and Horvitz (2004) DL: Litman et al. (1999) and Walker et al. (2000)
#ASRREJECTIONS	Number of ASR rejections up to this exchange e_n. $\sum_{i=1}^{n} x \begin{cases} 1 & \text{ASRRECOGNITIONSTATUS} = \text{REJECT}, \\ 0 & \text{otherwise} \end{cases}$ EL: Paek and Horvitz (2004) and Levin and Pieraccini (2006) DL: Kamm et al. (1998), Litman et al. (1999), Litman and Pan (1999), Walker et al. (2000) and ITU (2005)
{#}ASRREJECTIONS	Number of ASR rejections within the previous w turns prior to e_n, where w is the size of the window $\sum_{i=n-w}^{n} x \begin{cases} 1 & \text{ASRRECOGNITIONSTATUS} = \text{REJECT}, \\ 0 & \text{otherwise} \end{cases}$
#TIME-OUTS_ASRREJ	Number of time-out and ASR rejection events up to this exchange e_n $\sum_{i=1}^{n} x \begin{cases} 1 & \text{ASRRECOGNITIONSTATUS} \in \{\text{TIMEOUT}, \text{REJECT}\}, \\ 0 & \text{otherwise} \end{cases}$
{#}TIME-OUTS_ASRREJ	Number of time-out and ASR rejection events within the previous w turns prior to e_n, where w is the size of the window $\sum_{i=n-w}^{n} x \begin{cases} 1 & \text{ASRRECOGNITIONSTATUS} \in \{\text{TIMEOUT}, \text{REJECT}\}, \\ 0 & \text{otherwise} \end{cases}$
% TIME-OUTS _ASRREJ	Percentage of time-out and ASR rejection events in all previous exchanges: $\frac{1}{n}\sum_{i=1}^{n} x \begin{cases} 1 & \text{ASRRECOGNITIONSTATUS} \in \{\text{TIMEOUT}, \text{REJECT}\} \\ 0 & \text{otherwise} \end{cases}$ Exchanges in the beginning of the dialog where no ASR is active are replenished with "0"

(continued)

Table A.1 (continued)

Parameter	Description
BARGE-IN?	True if the user interrupted the system prompt, false otherwise ($\in \{true, false\}$)
# BARGE-INS	Number of barge-ins up to this exchange e_n $\sum_{i=1}^{n} x \begin{cases} 1 & \text{BARGEIN?} = \text{TRUE}, \\ 0 & \text{BARGEIN?} = \text{FALSE}. \end{cases}$ DL: Kamm et al. (1998), Litman et al. (1999), Litman and Pan (1999), Walker et al. (2000) and ITU (2005)
{#} BARGE-INS	Number of barge-ins within previous w turns prior to e_n, where w is the size of the window $\sum_{i=n-w}^{n} x \begin{cases} 1 & \text{BARGEIN?} = \text{TRUE}, \\ 0 & \text{BARGEIN?} = \text{FALSE}. \end{cases}$
% BARGE-INS	Percentage of barge-ins in all previous exchanges: $\frac{1}{n}\sum_{i=1}^{n} x \begin{cases} 1 & \text{BARGEIN?} = \text{TRUE}, \\ 0 & \text{BARGEIN?} = \text{FALSE}. \end{cases}$ DL: Litman et al. (1999) and Walker et al. (2000)
ASRCONFIDENCE	Confidence of the ASR module prepresenting the certainty of returning the correct ASR parse ($\in \mathbb{R}\{0\ldots1\}$) EL: Langkilde et al. (1999), Paek and Horvitz (2004), and Levin and Pieraccini (2006) DL: Kamm et al. (1998), Walker et al. (2000) and Litman and Pan (2002)
MEANASRCONFIDENCE	Average ASR confidence up to this exchange $\frac{1}{n}\sum_{i=1}^{n} x \begin{cases} ASRConfidence_i & \text{ASRREC.STATUS} \in \{\text{SUCC.}, \text{REJECT}\} \\ ASRConfidence_{cm} & \text{otherwise} \end{cases}$ Missing values are replenished with the average corpus-wide confidence $ASRConfidence_{cm}$ calculated on all "Success" and "Reject" turns. DL: Litman et al. (1999) and Litman and Pan (1999)
{MEAN} ASRCONFIDENCE	Average ASR confidence within previous w turns prior to e_n, where w is the size of the window $\frac{1}{n}\sum_{i=n-w}^{n} x \begin{cases} ASRConfidence_i & \text{ASRREC.STATUS} \in \{\text{SUCC.}, \text{REJECT}\} \\ ASRConfidence_{cm} & \text{otherwise} \end{cases}$ Missing values are replenished with the average corpus-wide confidence $ASRConfidence_{cm}$ calculated on all "Success" and "Reject" turns
UTD	Utterance Turn Duration of the user utterance in seconds EL: Langkilde et al. (1999), Walker et al. (2002), and Levin and Pieraccini (2006) DL: ITU (2005)
EXMO	Expected input modality by the system at current exchange $\in \{speech, dtmf, both, none\}$ EL: Langkilde et al. (1999) and Walker et al. (2002)
MODALITY	Input modality of the user for responding to a question $\in \{speech, dtmf\}$ EL: Langkilde et al. (1999), Walker et al. (2002), and Levin and Pieraccini (2006)

(continued)

Table A.1 (continued)

Parameter	Description
UnExMo?	User employed other modality than suggested by system prompt $\begin{cases} true & \text{Modality} \notin \text{ExMo} \\ false & \text{otherwise} \end{cases}$
#UnExMo	Number of unexpected modality usages up to this exchange e_n $\sum_{i=1}^{n} x \begin{cases} 1 & \text{Modality} \notin \text{ExMo} \\ 0 & \text{otherwise} \end{cases}$
{#}UnExMo	Number of unexpected modality usages within previous w turns prior to e_n, where w is the size of the window $\sum_{i=n-w}^{n} x \begin{cases} 1 & \text{Modality} \notin \text{ExMo} \\ 0 & \text{otherwise} \end{cases}$
%UnExMo	Percentage of unexpected modality usage up to this exchange $\frac{1}{n}\sum_{i=1}^{n} x \begin{cases} 1 & \text{Modality} \notin \text{ExMo} \\ 0 & \text{otherwise} \end{cases}$
WPUT	Number of words returned in the parse EL: Langkilde et al. (1999), Walker et al. (2002), and Levin and Pieraccini (2006) DL: ITU (2005)

Related work using similar parameters on the exchange level (EL) and on the dialog level (DL) is given respectively

A.2 Interaction Parameters Derived from Language Understanding

Table A.2 Interaction parameters derived from the Language Understanding (LU) module for employment on the exchange level

Parameter	Description
SemanticParse	Semantic parse of the caller utterance as returned by the activated grammar EL: Litman et al. (1999)
HelpRequest?	Current turn is a (from the system recognized) help request, i.e., the user asks for more details ($\in \{true, false\}$)
#HelpRequests	Number of help requests up to this exchange e_n $\sum_{i=1}^{n} x \begin{cases} 1 & \text{HelpRequest?} = \text{TRUE}, \\ 0 & \text{otherwise} \end{cases}$ EL: Paek and Horvitz (2004) DL: Kamm et al. (1998), Litman and Pan (1999), ITU (2005), and Hajdinjak and Mihelic (2006)
{#}HelpRequests	Number of help requests within previous w turns prior to e_n, where w is the size of the window $\sum_{i=n-w}^{n} x \begin{cases} 1 & \text{HelpRequest?} = \text{TRUE}, \\ 0 & \text{otherwise} \end{cases}$

(continued)

Table A.2 (continued)

Parameter	Description
% Help Requests	Percentage of help-requests in all previous exchanges: $\frac{1}{n}\sum_{i=1}^{n} x \begin{cases} 1 & \text{HelpRequest?} = \text{TRUE} \\ 0 & \text{otherwise} \end{cases}$ EL: Paek and Horvitz (2004) DL: Litman et al. (1999) and Hajdinjak and Mihelic (2006)
Operator Request?	Current turn is a (from the system recognized) request for an operator, i.e. the user opts out ($\in \{true, false\}$). EL: Paek and Horvitz (2004)
# Operator Requests	Number of operator requests up to this exchange e_n $\sum_{i=1}^{n} x \begin{cases} 1 & \text{OperatorRequest?} = \text{TRUE}, \\ 0 & \text{otherwise} \end{cases}$ EL: Paek and Horvitz (2004)
{#}Operator Requests	Number of operator requests within previous w turns prior to e_n, where w is the size of the window $\sum_{i=n-w}^{n} x \begin{cases} 1 & \text{OperatorRequest?} = \text{TRUE}, \\ 0 & \text{otherwise} \end{cases}$
% Operator requests	Percentage of operator requests in all previous exchanges: $\frac{1}{n}\sum_{i=1}^{n} x \begin{cases} 1 & \text{Operator?} = \text{TRUE} \\ 0 & \text{otherwise} \end{cases}$ EL: Paek and Horvitz (2004)

Related work using similar parameters on the exchange level (EL) and on the dialog level (DL) is given respectively

A.3 Interaction Parameters Derived from Dialog Management

Table A.3 Interaction parameters derived from the Dialog Management (DM) module for employment on the exchange level

Parameter	Description
Activity	The name of the activity that was performed by the system consisting of an identifier for the question or statement. Activities of different dialog systems are according to the flowchart design respectively, e.g., a bus information service may have activities like "query.arrival_place", "query.travel_time", "confirm_okay", etc. The names are determined by the system designer EL: Langkilde et al. (1999) and Walker et al. (2002)
ActivityTrigram	Sequence of the current activity plus the two previous activities. This feature models the history of activities
ActivityType	Type of activities $\in$ {"announcement","question","confirmation","wait"}
DD	Dialog duration in seconds up to this exchange EL:Langkilde et al. (1999) and Walker et al. (2002) DL: Walker et al. (2000) and ITU (2005)

(continued)

Table A.3 (continued)

Parameter	Description
PROMPT	System prompt of the automated agent prior to recording user input EL: Langkilde et al. (1999) and Walker et al. (2002)
ROLE INDEX	In a dialog module activity, the number of tries to elicit a desired response from the user
ROLE NAME	The function of this system turn $\in$ {''collection'',''confirmation''}
REPROMPT?	Current turn is a *reprompt* $\in \{true, false\}$ EL: Langkilde et al. (1999) and Walker et al. (2002)
# REPROMPT	Number of reprompts up to this exchange e_n $\sum_{i=1}^{n} x \begin{cases} 1 & \text{Reprompt?}=true, \\ 0 & \text{otherwise} \end{cases}$ EL: Langkilde et al. (1999), Walker et al. (2002), and Levin and Pieraccini (2006)
{#}REPROMPT	Number of reprompts within previous w turns prior to e_n, where w is the size of the window $\sum_{i=n-w}^{n} x \begin{cases} 1 & \text{REPROMPT?} = \text{TRUE}, \\ 0 & \text{otherwise} \end{cases}$
% REPROMPT	Percentage of reprompts in all previous exchanges: $\frac{1}{n} \sum_{i=1}^{n} x \begin{cases} 1 & \text{REPROMPT?} = \text{TRUE} \\ 0 & \text{otherwise} \end{cases}$ EL: Langkilde et al. (1999) and Walker et al. (2002)
LOOP NAME	Given caller speech input, we need to try and recognize the semantic meaning. The first time of try is indicated with a value of 'Initial'. If the system could not turn a parse then we have to reprompt ("Retry1" or "Timeout1"). Similar for if the caller asks for help or a repetition of the prompt
EXCHANGES, #SYSTEMTURNS, #USERTURNS	Number of exchanges and system/user turns up to this exchange. The number of turns may differ from the number of exchanges since exchanges may also exist of a single user or system turn EL: Langkilde et al. (1999), Walker et al. (2002) and Levin and Pieraccini (2006) DL: Litman and Pan (1999), ITU (2005) and Möller et al. 2008
# SYSTEM QUESTIONS	Number of system questions up to this exchange DL: ITU (2005)
{#} SYSTEM QUESTIONS	Number of system questions in the current sliding window
WPST	Words per system turn EL: Litman et al. (1999) DL: ITU (2005)

Related work using similar parameters on the exchange level (EL) and on the dialog level (DL) is given respectively

Appendix B
Detailed Results for Emotion Recognition, Interaction Quality, and Task Success

B.1 Emotion Recognition

Table B.1 Distribution of anger in the utterances of the SpeechCycle Broadband Agent Corpus SC-BAC, separated by gender

Emotion	Quantity	Percentage	Callers	Average call length (in exchanges)
All utterances				
Other	20500	90.27	1872	13.30
Annoyed	770	3.35	393	13.44
Hot Anger	161	0.72	92	10.75
Garbage	1167	5.15	572	17.93
Undefined	113	0.51	90	13.32
Female speakers				
Other	10102	94.86	962	13.23
Annoyed	402	3.78	215	12.96
Hot Anger	70	0.66	40	8.70
Undefined	75	0.70	53	12.69
Male speakers				
Other	10346	95.42	925	13.36
Annoyed	368	3.39	178	13.97
Hot Anger	91	0.84	52	12.32
Undefined	38	0.35	37	14.55

Labels are a result of majority voting from three raters. Where no majority voting was possible, since all three raters differed, the sample has been marked as "undefined"

A. Schmitt and W. Minker, *Towards Adaptive Spoken Dialog Systems*,
DOI: 10.1007/978-1-4614-4593-7,

Table B.2 Human performance on SC-BAC: 17 raters respectively annotated 33 *non-angry*, 33 *slightly angry*, and 33 *very angry* utterances from the corpus

	Accuracy (%)	NonAnger		SlightAnger		StrongAnger		NonAnger	SlightAnger	StrongAnger	
		Precision (%)	Recall (%)	Precision (%)	Recall (%)	Precision (%)	Recall (%)	*F*	*F*	*F*	*fl*
	50.5	55.2	97.0	31.3	30.3	88.9	24.2	0.703	0.308	0.381	0.464
	67.7	69.8	90.9	51.4	57.6	94.7	54.6	0.789	0.543	0.692	0.675
	71.7	75.0	81.8	58.3	63.6	85.2	69.7	0.783	0.609	0.767	0.719
	58.6	65.8	75.8	40.0	48.5	81.0	51.5	0.704	0.438	0.630	0.591
	53.5	58.8	90.9	35.1	39.4	90.9	30.3	0.714	0.371	0.455	0.513
	69.7	78.4	87.9	54.1	60.6	80.0	60.6	0.829	0.571	0.690	0.697
	61.6	60.8	93.9	44.4	36.4	85.7	54.6	0.738	0.400	0.667	0.602
	64.6	73.7	84.9	48.4	45.5	70.0	63.6	0.789	0.469	0.667	0.641
	65.7	71.4	90.9	48.8	60.6	93.8	45.5	0.800	0.541	0.612	0.651
	63.6	66.7	90.9	47.1	48.5	85.0	51.5	0.769	0.478	0.642	0.629
	60.6	64.4	87.9	44.4	48.5	83.3	45.5	0.744	0.464	0.588	0.598
	66.7	72.4	63.6	50.0	69.7	91.7	66.7	0.677	0.582	0.772	0.677
	61.6	57.4	93.9	46.2	36.4	94.7	54.6	0.713	0.407	0.692	0.604
	61.6	60.8	93.9	44.4	36.4	85.7	54.6	0.738	0.400	0.667	0.602
	71.7	66.7	90.9	64.0	48.5	86.2	75.8	0.769	0.552	0.806	0.709
	61.6	62.5	90.9	45.5	45.5	88.9	48.5	0.741	0.455	0.627	0.608
	58.6	60.9	84.9	42.1	48.5	93.3	42.4	0.709	0.451	0.583	0.581
mean	0.629	65.9	87.7	46.8	48.5	87.0	52.6	0.748	0.473	0.643	0.621
std	0.058	6.7	8.1	7.8	11.0	6.4	13.1	0.042	0.083	0.106	0.067

B.1.1 Evolvement of Anger: History of Anger in SC-BAC

Table B.3 Distribution of anger in SC-BAC

		NonAnger	Garbage	SlightAnger	StrongAnger
NonAnger observed	n − 2	93.0	4.3	2.5	0.2
	n − 1	93.5	4.0	2.3	0.2
SlightAnger observed	n − 2	78.5	6.5	12.5	2.6
	n − 1	65.8	6.8	23.7	3.7
StrongAnger observed	n − 2	45.4	10.8	20.8	23.1
	n − 1	36.9	5.0	19.9	38.3

If an emotion ∈ NoAnger, SlightAnger, StrongAnger is observed at turn n, then the previous two turns $n-1$ and $n-2$ contained the following emotional state

B.1.2 Language Comparison

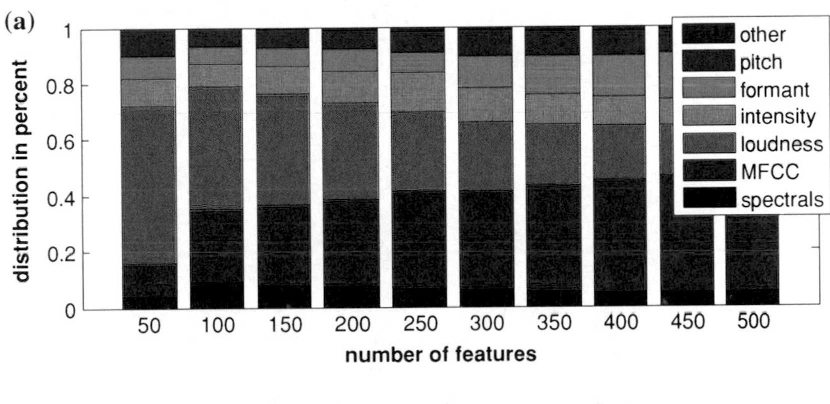

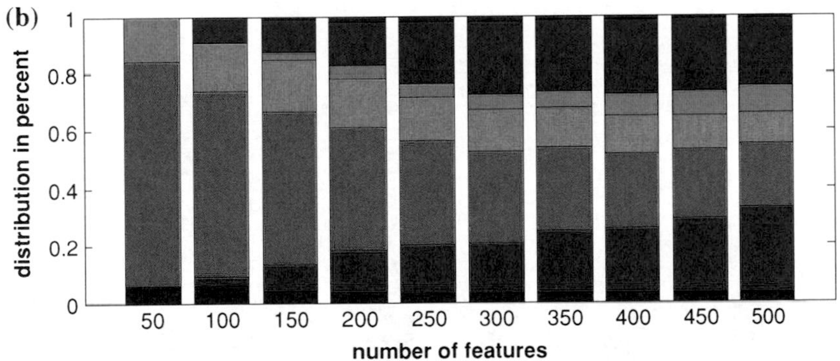

Fig. B.1 Percentage of features from P_{ext} within each feature group when considering the 50–500 top ranked features according to IGR. Number of features in total = 1450. Loudness features dominate in the English SC-BAC data set. For details refer to Polzehl et al. (2009a). **a** MOB corpus (German), **b** SC-BAC corpus (US-English)

B.2 IQ Annotations

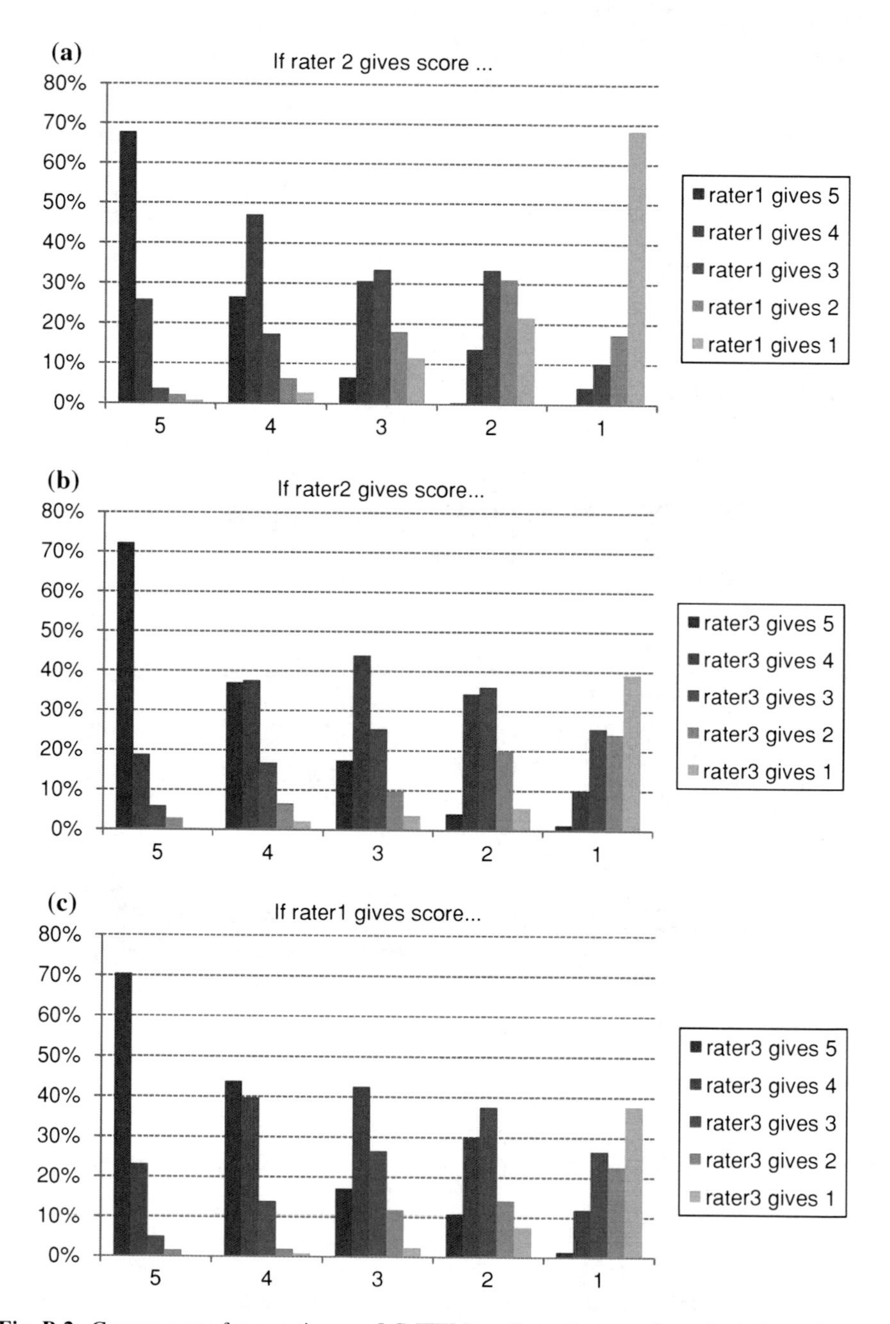

Fig. B.2 Congruency of annotations on LG-FIELD. **a** Rater 2 versus Rater 1, **b** Rater 2 versus Rater 3, **c** Rater 1 versus Rater 3

B.1 Task Descriptions for LG-LAB User Study

The tasks that have been set to the users for the Interaction Quality study are depicted in the following (Figs. B.1 and B.2).

- *Task 1* (answerable)You are on a business trip to Pittsburgh, Pennsylvania and you have just arrived at the airport.

The hotel where you are going to stay is in downtown Pittsburgh. You are planning to go there by bus, but you do not have any schedule information about the bus system there.

So you decide to call the Let's Go Bus Information System to obtain this information. The first meeting is going to start soon, so you want to check in your hotel as quickly as possible.

Please try to obtain schedule information about a bus route from *the airport* to *Downtown Pittsburgh* for the *next bus* and the *one after that.*

- *Task 2* (answerable)Assume you are a student at the University of Pittsburgh.

Today, your lecture courses will finish shortly before 4 p.m. . Unfortunately one of your fellow students is very sick and now hospitalized at Mercy Hospital. You are planning a sick bed visit afterwards.

Please try to obtain schedule information for a bus running from the *University of Pittsburgh* to *Mercy Hospital* at *4 p.m.* as well as *the previous bus.*

- *Task 3* (answerable)This morning you went shopping in downtown Pittsburgh. Because of the daily traffic jam you parked your car at the Park and Ride area at Bell Station and continued by bus.

You finally got all your purchases in Liberty Avenue and want to go back home at 10 a.m. .

Please try to obtain schedule information for a bus running from *Liberty Avenue* to *Bell Station* at *10 a.m.*.

- *Task 4* (non answerable)This evening you are invited for dinner near the North Side of Pittsburgh. You know that you have to take Bus 500 but you are not sure about the time it will leave.

Please try to obtain schedule information for *Bus 500* from the *West View* to the *North Side* at *5 p.m.* this afternoon.

- *Task 5* (answerable)You have planned to walk back from Hamilton to downtown Pittsburgh. As it unfortunately begins to rain you decide to take a Bus to return at once.

Please try to obtain schedule information for a Bus from *Hamilton* to *Downtown Pittsburghimmediately*.

B.4 Age Recognition

Analyzing the task completion rate of SC-BAC we found out that non-senior callers had a 33 % higher task completion rate than senior callers. An adaption of the dialog to elderly users could potentially raise their task completion rate, e.g., by introducing more explicit confirmations or in case of additional problems an immediate transfer to an operator who can help out. Other application scenarios for a distinction between senior and non-senior users are:

- A shifting of the acoustic ASR models to models that are especially trained on aged voices to raise the recognizer's accuracy.
- Self-service applications that employ advertisements tailored to the specific user group while the caller is on hold for a live operator.

Estimating speaker age becomes increasingly difficult the shorter the utterances are. We could observe this phenomenon during the rating process. The raters were presented the calls from SC-BAC in descending order, i.e., the calls at the beginning of the rating process contained longer utterances than at the end of the process. The growing uncertainty can be observed in Fig. B.3. Figure B.3a presents the number of calls labeled as non-seniors. It is interesting to note that regardless of the utterance length the raters constantly labeled calls as non-seniors. Obviously the decision on rating utterances as "non-senior" was not affected by the length. Rating senior speakers seems to be a less trivial task. With decreasing utterance length, the raters labeled less calls as 'senior' and chose 'unsure' instead. Feature-specific classification results are depicted in Fig. B.4. For details on age recognition refer to Schmitt et al. (2010d) (Tables A.1, A.2 and A.3).

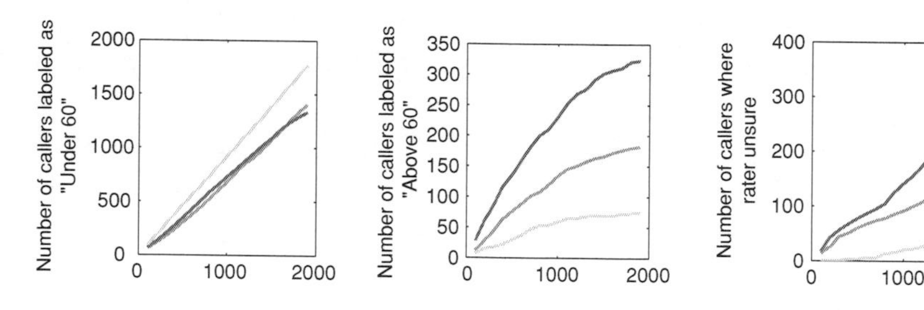

Fig. B.3 Rising uncertainty in manual age annotation with decreasing sample length. Three Raters annotated age of 1,911 callers and choose between ">60", "≤ 60" and "unsure". *First plot* Number of calls the rater annotated with '≤ 60'. *Second plot* Number of calls each rater annotated with '>60'. *Third plot* Number of calls where the raters were unsure. It should be noted that the length of the presented sample decreases. As can be seen, with decreasing utterance length the uncertainty increases and the raters favor for 'unsure' instead of '>60'

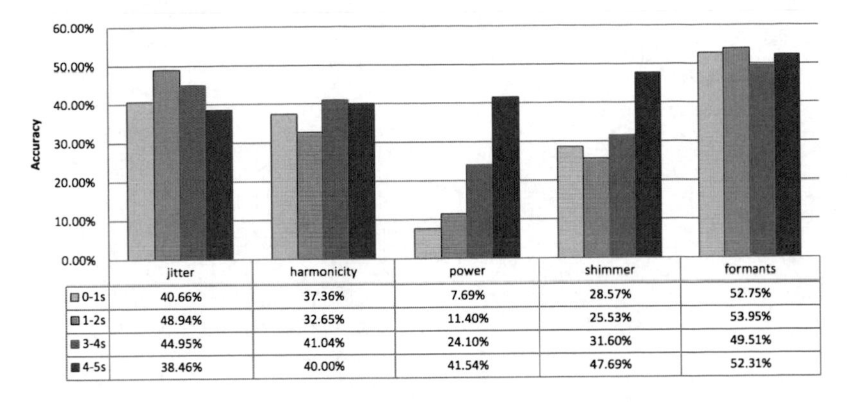

	jitter	harmonicity	power	shimmer	formants
0-1s	40.66%	37.36%	7.69%	28.57%	52.75%
1-2s	48.94%	32.65%	11.40%	25.53%	53.95%
3-4s	44.95%	41.04%	24.10%	31.60%	49.51%
4-5s	38.46%	40.00%	41.54%	47.69%	52.31%

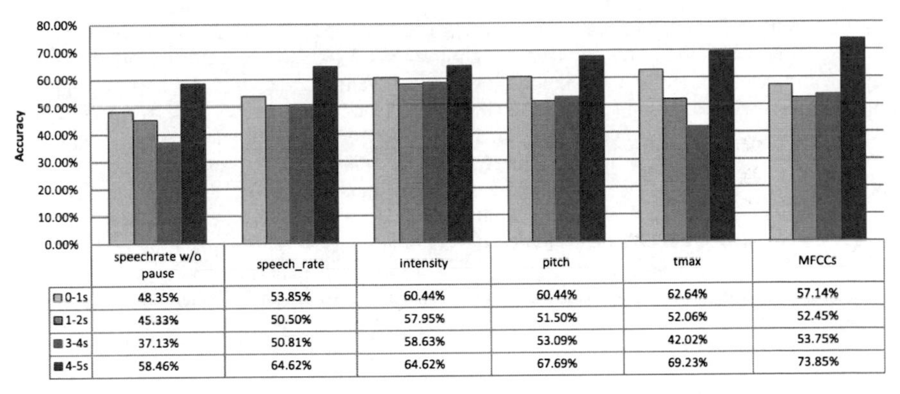

	speechrate w/o pause	speech_rate	intensity	pitch	tmax	MFCCs
0-1s	48.35%	53.85%	60.44%	60.44%	62.64%	57.14%
1-2s	45.33%	50.50%	57.95%	51.50%	52.06%	52.45%
3-4s	37.13%	50.81%	58.63%	53.09%	42.02%	53.75%
4-5s	58.46%	64.62%	64.62%	67.69%	69.23%	73.85%

Fig. B.4 Percentage of correctly classified utterances within selected feature groups analyzed according to the utterance length. Jitter, harmonicity, and formants do not gain performance with increasing utterance length. In contrast, substantial increase can be observed when the utterance lasts at least 4 s which affects particularly power, shimmer, intensity, pitch, tmax, and MFCCs

B.5 Example Architecture for Realizing Speaker-Awareness in VoiceXML-Based SDS

This section presents a speaker classification architecture for VoiceXML-based applications.

VoiceXML 2.1 is, without doubt, the established standard for developing Spoken Language Dialog Systems (SLDS) for telephone applications. While the development of state-of-the-art Voice User Interfaces (VUIs) has been strongly facilitated with VoiceXML, next-generation user interfaces can only be built with limitations. Indeed, next-generation techniques, such as mixed-initiative (VoiceXML 2.1) and multimodality (X+V, XHTML+Voice) have been rudimentarily considered, however, important elements have been left disregarded. Current systems are able to understand what the user says and what he means, but lack further information about the user himself and the group he belongs to (Tables B.1, B.2 and B.3).

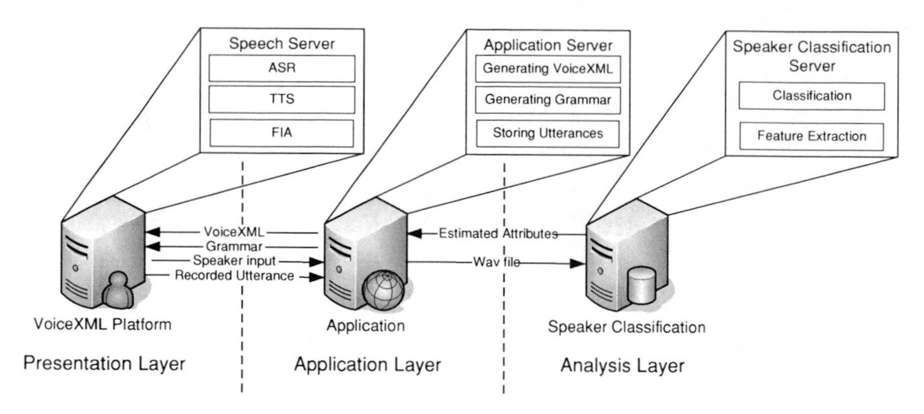

Fig. B.5 A three-layered architecture for speaker classification. The Presentation Layer consists of ASR, TTS, and the Form Interpretation Algorithm (see VoiceXML specification). We use the same naming conventions as in common three-tier software architectures since speech applications are in many aspects similar to regular GUI applications. The Presentation Layer can be seen as a client that receives information regarding dialog flow (VoiceXML) and grammar (grxml or GSL) from the second layer, the Application Layer. During dialog execution it sends the application-dependent speaker input and the last user utterance back to the Application Layer. The Application Layer both serves as a kind of middleware between the Presentation and the Analysis Layer and as a server that provides the application itself. This application may generate the VoiceXML encoded dialog for a travel agency, a technical support automated agent, a music store, etc. The task of the Analysis Layer is to classify user-specific properties, such as age, gender, emotional state, etc. Based on the knowledge generated by the Analysis Layer, the Application Layer may generate personalized and adapted prompts

Although the current draft for the next version of the standard, VoiceXML 3.0, envisions *speaker verification* and *speaker identification* it does not allow for *speaker classification*. It should be noted that there are significant differences: verification determines whether you are really the one you are claiming to be, e.g., for banking applications, whereas identification determines who you are from a distinct set of users, i.e., as an individual, e.g., for intelligent home applications. In contrast to this, speaker classification assigns the speaker to a certain group of users, such as male versus female, teens versus middle-aged adults versus elderly, angry versus neutral, non-native speakers versus native speakers, etc. This additional information can contribute to an adapted, personalized conversation between user and system and would let the system appear more "intelligent" and more or less on equal footing with the user compared with today's systems.

For the VoiceXML platform, speaker classification would be a rather easy and not too costly task since most relevant features such as MFCC coefficients (Mel Frequency Cepstral Coefficients) are calculated anyway for Automatic Speech Recognition (ASR). Models could be adequately generalized and trained to work with a broad range of applications, telephone channels, and background noises as it is currently the case for ASR. However, neither the VoiceXML working group nor the platform manufacturers seem to push forward in that direction.

```
<form id="listeningLoop">
  <property name="recordutterance" value="true"/>
    <field name="waitForCommand">
      <grammar src="http://applicationServerURL.org/grammar.grxml"/>
      <filled>
        <var name="userRecording" expr="waitForCommand$.recording"/>
        <data name="xmlAnswer" method="POST" enctype="multipart/form-data"
        namelist="userRecording" srcexpr="'http://applicationServerURL.org/Receive'"/>
        <script>
          <![CDATA[
            try {
              document.personalizedPrompt=xmlAnswer.documentElement.getElementsBy
              TagName("personalizedPrompt").item(0).firstChild.data;
            } catch(e){}
          ]]>
        <script/>
        <prompt>document.personalizedPrompt</prompt>
        :
```

Fig. B.6 The essential part of a VoiceXML snippet used for speaker classification. Important elements are the "recordutterance" attribute, the data- and the script tag. In this example, a keyword activation function for an Intelligent Environment, such as an intelligent home, is demonstrated. The dialog system is kept in a listening mode until a "magic word" is recognized. The property "recordutterance" allows to store the last user utterance in a platform variable that can be accessed by "name_of_field$.recording". The data tag is used to pass the utterance encoded as a "wav" audio file to the Application Layer. As response, the data tag receives an XML file containing the Application Layer's personalized prompt based on the Analysis Layer's classification result. The personalized system utterance is extracted by a line of Java-Script code encapsulated in the script tag. Finally the prompt tag presents this utterance to the user. With the same technique, not only can a prompt be encapsulated within the XML answer, but a URL pointing to a personalized VoiceXML dialog

The proposed architecture, see Fig. B.5, aims to provide a workaround that enables speaker-awareness in today's systems. Our analysis component receives user utterances from the VoiceXML platform, performs feature extraction, and classifies speaker characteristics such as age, gender, and emotional state. This additional information about the speaker can be employed to adapt system prompts and the dialog strategy to specific user groups. For integrating speaker-awareness into the VoiceXML dialect, a dialog script must be adapted as depicted in Fig. B.5. Details may be found in Schmitt et al. (2009a) (Fig. B.6).

References

Abdennadher, S., Aly, M., Bühler, D., Minker, W., & Pittermann, J. (2007). BECAM tool—a semi-automatic tool for bootstrapping emotion corpus annotation and management. In *Proceedings of European Conference on Speech and Language Processing (EUROSPEECH)* (pp. 946–949). Belgium: Antwerp.

Abdulla, W. H., & Kasabov, N. K. (2001). Improving speech recognition performance through gender separation. In *Proceedings of Artificial Neural Networks and Expert Systems International Conference (ANNES)* (pp. 218–222).

Acomb, K., Bloom, J., Dayanidhi, K., Hunter, P., Krogh, P., Levin, E. et al. (2007). Technical support dialog systems:issues, problems, and solutions. In *Proceedings of the Workshop on Bridging the Gap: Academic and Industrial Research in Dialog Technologies* (pp. 25–31). Rochester, NY: Association for Computational Linguistics.

Ai, H., Litman, D. J., Forbes-riley, K., Rotaru, M., Tetreault, J., & Pur, A. (2006). Using system and user performance features to improve emotion detection in spoken tutoring dialogs. In *Proceedings of International Conference on Speech and Language Processing (ICSLP)* (pp. 797–800).

Albalate, A., & Minker, W. (2011). *Rapid design*. ISTE/Wiley, London (UK): Adaptation and Improvement of Statistical Spoken Language Understanding Systems.

Alexander, C., Ishikawa, S., & Silverstein, M. (1977). *A pattern language: Towns, buildings, construction*. New York: Oxford University Press.

Allen, J. (1995). *Natural language understanding* (2nd ed.). Redwood City, CA, USA: Benjamin-Cummings Publishing Co., Inc.

Ang, J., Dhillon, R., Krupski, A., Shriberg, E., & Stolcke, A. (2002). Prosody-based automatic detection of annoyance and frustration in human-computer dialog. In *Proceedings of International Conference on Speech and Language Processing (ICSLP)* (pp. 2037–2040).

Arnold, M. B. (1960). *Emotions and personality*. New York: Columbia University Press.

Barras, C., Geoffrois, E., Wu, Z., & Liberman, M. (2001). Transcriber: Development and use of a tool for assisting speech corpora production. *Speech Communication,33*, 5–22.

Bartholomew, D. J. (2002). *The analysis and interpretation of multivariate data for social scientists*. Texts in statistical science. New York: Chapman& Hall/CRC.

Batliner, A., Fischer, K., Huber, R., Spilker, J., & Nöth, E. (2000). Desperately seeking emotions: Actors, wizards, and human beings. In R. Cowie, E. Douglas-Cowie & M. Schröder (Eds.), *Proceedings of ISCA Workshop on Speech and, Emotion* (pp. 195–200).

Batliner, A., Hacker, C., Steidl, S., Nöth, E., D'Arcy, S., Russell, M. et al. (2004). "You stupid tin box"—children interacting with the AIBO robot: A cross-linguistic emotional speech corpus. In ELRA (Ed.), *Proceedings of the 4th International Conference of Language Resources and Evaluation LREC 2004*.

A. Schmitt and W. Minker, *Towards Adaptive Spoken Dialog Systems*,
DOI: 10.1007/978-1-4614-4593-7,

Baum, L. E., Petrie, T., Soules, G., & Weiss, N. (1970). A maximization technique occurring in the statistical analysis of probabilistic functions of Markov chains. *The Annals of Mathematical Statistics,41*(1), 164–171.

Bennett, K. P., & Campbell, C. (2000). Support vector machines: Hype or hallelujah? *Journal of SIGKDD Explorations,2*(2), 1–13.

Bernsen, N. O., & Dybkjaer, L. (1997). *Designing interactive speech systems: From first ideas to user testing* (1st ed.). New York, Secaucus: Springer.

Bernsen, N. O., Dybkjaer, L., & Kolodnytsky, M. (2002). The nite workbench—a tool for annotation of natural interactivity and multimodal data. In *Proceedings of the 3rd International Conference on Language Resources and Evaluation (LREC 2002)* (pp. 43–49). Spain: Las Palmas.

Bertrand, G., Heinroth, T., & Schmitt, A. (2009). Chad—constraint handling architecture for dialoguemanagement. In *5th International Conference on Intelligent Environments (IE' 09), volume 2 of Ambient Intelligence and Smart Environments* (pp. 50–56). Amsterdam: IOS Press.

Bezold, M., & Minker, W. (2011). *Adaptive multimodal interactive systems*. Boston (USA): Springer.

Bitouk, D., Verma, R., & Nenkova, A. (2010). Class-level spectral features for emotion recognition. *Speech Communication,52*(7–8), 613–625.

Black, A., & Eskenazi, M. (2009). The spoken dialogue challenge. In *Proceedings of the SIGDIAL 2009 Conference* (pp. 337–340). London, UK: Association for Computational Linguistics.

Bocklet, T., Maier, A., Bauer, J., Burkhardt, F., & Nöth, E. (2008). Age and gender recognition for telephone applications based on gmm supervectors and support vector machines. In I. C. S. Press (Ed.), *Proceedings of the International Conference on Acoustics, Speech, and Signal Processing (ICASSP)* (Vol. 1, pp. 1605–1608).

Boersma, P., & Weenink, D. (2009). Praat: Doing phonetics by computer. Computer program (version 5.1.41).

Bohus, D., Raux, A., Harris, T. K., Eskenazi, M., & Rudnicky, A. I. (2007). Olympus: An open-source framework for conversational spoken language interface research. In *Proceedings of the workshop on bridging the gap: Academic and industrial research in dialog technologies, NAACL-HLT-Dialog '07* (pp. 32–39). Prague: Association for Computational Linguistics.

Bosma, W., & André, E. (2004). Exploiting emotions to disambiguate dialogue acts. In *IUI '04: Proceedings of the 9th international conference on Intelligent user interfaces* (pp. 85–92). New York, NY, USA: ACM.

Bray, T., Paoli, J., Sperberg-McQueen, C. M., Maler, E., & Yergeau, F. (2004). *Extensible markup language (XML) 1.0* (3rd ed.). W3C recommendation. Technical report, W3C.

Breiman, L. (1996). Bagging predictors. *Machine Learning,24*, 123–140.

Burkhardt, F., Ajmera, J., Englert, R., Stegmann, J., & Burleson, W. (2006). Detecting anger in automated voice portal dialogs. In *Proceedings of International Conference on Speech and Language Processing (ICSLP) 2006*.

Burkhardt, F., Huber, R., & Batliner, A. (2007a). *Application of speaker classification in human machine dialog systems* (pp. 174–179). Berlin, Heidelberg: Springer. Speaker classification 1: Fundamentals, features, and methods edition.

Burkhardt, F., Huber, R., & Stegmann, J. (2008). Advances in anger detection with real life data. In *Proceedings of 19*. Frankfurt: Konferenz Elektronische Sprachsignalverarbeitung (ESSV).

Burkhardt, F., van Ballegooy, M., Englert, R., & Huber, R. (2005a). An emotion-aware voice portal. In *Proceedings of Electronic Speech Signal Processing ESSP*.

Burkhardt, F., Metze, F., & Stegmann, J. (2007b). *Speaker Classification for Next Generation Voice Dialog Systems*, chapter Speaker Classification for Next Generation Voice Dialog Systems. Advances in Digital Speech Transmission. New York: Wiley.

Burkhardt, F., Polzehl, T., Stegmann, J., Metze, F., & Huber, R. (2009). Detecting real life anger. In *Proceedings of the International Conference on Acoustics, Speech, and Signal Processing (ICASSP)*.

Burkhardt, F., Rolfes, M., Sendlmeier, W., & Weiss, B. (2005b). A database of german emotional speech. In *Proceedings of the International Conference on Speech and Language Processing (ICSLP) Interspeech 2005* (pp. 1517–1520). New York: ISCA.

Campbell, N., Devillers, L., Douglas-Cowie, E., Auberge, V., Batliner, A., & Tao, J. (2006). Resources for the processing of affect in interactions. In *Proceedings of the International Conference on Language Resources and Evaluation (LREC)*. Amsterdam: ELRA.

Carletta, J. (1996). Assessing agreement on classification tasks: The kappa statistic. *Computational Linguistics Linguistics, 22*(2), 249–254.

Carletta, J., Evert, S., Heid, U., & Kilgour, J. (2005). The nite xml toolkit: Data model and query language. *Language Resources and Evaluation,39*, 313–334. doi:10.1007/s10579-006-9001-9.

Carletta, J., Evert, S., Heid, U., Kilgour, J., Robertson, J., & Voormann, H. (2003). The nite xml toolkit: Flexible annotation for multimodal language data. *Behavior Research Methods,35*, 353–363. doi:10.3758/BF03195511.

Cohen, J. (1960). A coefficient of agreement for nominal scales. *In Educational and Psychological Measurement,20*, 37–46.

Cohen, M. H., Giangola, J. P., & Balogh, J. (2004). *Voice user interface design*. Redwood City, CA, USA: Addison Wesley Longman Publishing Co., Inc.

Cohen, P. R. (1995a). *Empirical methods for artificial intelligence*. Cambridge, MA, USA: MIT Press.

Cohen, W. W. (1995b). Fast effective rule induction. In *Proceedings of the 12th International Conference on Machine Learning*.

Cohen, W. W. (1996). Learning trees and rules with set-valued features. In *Proceedings of the thirteenth national conference on Artificial intelligence—Volume 1, AAAI'96* (pp. 709–716). Cambridge: AAAI Press.

Cohen, W. W., & Singer, Y. (1999). A simple, fast, and effective rule learner. In *Proceedings of the Sixteenth National Conference on Artificial Intelligence* (pp. 335–342). Cambridge: AAAI Press.

Cornelius, R. (1996). *The science of emotion: Research and tradition in the psychology of emotions*. Upper Saddle River, London: Prentice Hall.

Cornelius, R. (2000). Theoretical approaches to emotion. In *Proceedings of the ISCA Workshop on Speech and Emotion* (pp. 3–10). ISCA Tutorial and Research Workshop (ITRW).

Cortes, C., & Vapnik, V. (1995). Support-vector networks. *Machine Learning,20*, 273–297.

Cowie, R., Douglas-Cowie, E., Tsapatsoulis, N., Votsis, G., Kollias, S., Fellenz, W., et al. (2001). Emotion recognition in human-computer interaction. *IEEE Signal Processing Magazine,18*(1), 32–80.

Crawley, M. J. (2007). *The R book* (1st ed.). Chichester: Wiley.

Danieli, M., & Gerbino, E. (1995). Metrics for evaluating dialogue strategies in a spoken language system. In *Working Notes AAAI Spring Symposium on Empirical Methods in Discourse Interpretation and Generation* (pp. 34–39).

Darwin, C. (1872). *The expression of the emotions in man and animals*. London: John Murray.

Davies, K., Biddulph, R., & Balashek, S. (1952). Automatic speech recognition of spoken digits. *Journal of the Acoustical Society of America,24*(6), 637–642.

Davies, M., & Fleiss, J. (1982). Measuring agreement for multinomial data. *Biometrics,38*, 1047–1051.

Devillers, L., Lamel, L., & Vasilescu, I. (2003). Emotion detection in task-oriented spoken dialogues. In *Proceedings of the 2003 International Conference on Multimedia and Expo—Volume 3 (ICME '03)—Volume 03, ICME '03* (pp. 549–552). Washington, DC, USA: IEEE Computer Society.

Devillers, L., & Vidrascu, L. (2006). Real-life emotions detection with lexical and paralinguistic cues on human-human call center dialogs. In *Proceedings of the International Conference on Speech and Language Processing (ICSLP)*

Doll, W. J., & Torkzadeh, G. (1991). The measurement of end-user computing satisfaction: Theoretical and methodological issues. *MIS Quarterly, 15*, 5–10.

Domingos, P. (1999). Metacost: A general method for making classifiers cost-sensitive. In *Proceedings of the Fifth International Conference on Knowledge Discovery and Data Mining* (pp. 155–164). New York: ACM Press.
Douglas-Cowie, E., Cowie, R., & Schröder, M. (2000). *A new emotion database: Considerations, sources and scope* (pp. 39–44). International Speech Communication Association.
Duda, R. O., Hart, P. E., & Stork, D. G. (2001). *Pattern classification* (2nd ed.). New York: Wiley-Interscience.
Dybkjaer, L., Hemsen, H., & Minker, W. (2007). *Introduction, volume 37 of text, speech and language technology, pp. vi–xxiii.* Dordrecht (The Netherlands): Springer. Evaluation of Text and Speech Systems edition.
Dyer, R. J. T. (2005). *MySQL in a nutshell (In a nutshell (O'Reilly)).* Sebastopol: O'Reilly Media, Inc.
Eckert, W., Levin, E., & Pieraccini, R. (1998). Automatic evaluation of spoken dialogue systems. In *TWLT13: Formal semantics and pragmatics of dialogue* (pp. 99–110).
Ekman, P. (1992). An argument for basic emotions. *Cognition & Emotion,6*(3), 169–200.
Enberg, I. S., & Hansen, A. V. (1996). *Documentation of the danish emotional speech database.* Technical report: Aalborg University, Denmark.
Engelbrecht, K.-P., Gödde, F., Hartard, F., Ketabdar, H., & Möller, S. (2009). Modeling user satisfaction with hidden markov model. In *SIGDIAL '09: Proceedings of the SIGDIAL 2009 Conference* (pp. 170–177). Morristown, NJ, USA: Association for Computational Linguistics.
Engelbrecht, K.-P., Kühnel, C., & Möller, S. (2008). Weighting the coefficients in paradise models to increase their generalizability. In *Proceedings of the 4th IEEE tutorial and research workshop on Perception and Interactive Technologies for Speech-Based Systems: Perception in Multimodal Dialogue Systems, PIT '08* (pp. 289–292), Berlin, Heidelberg: Springer.
Eppinger, B., & Herter, E. (1993). *Sprachverarbeitung.* München: Carl-Hanser-Verlag.
Evanini, K., Hunter, P., Liscombe, J., Suendermann, D., Dayanidhi, K., & Pieraccini, R. (2008). Caller experience: A method for evaluating dialog systems and its automatic prediction. In *Spoken Language Technology Workshop, 2008. SLT 2008. IEEE* (pp. 129–132).
Fastl, H., & Zwicker, E. (2005). *Psychoacoustics: Facts and models* (3rd ed.). Berlin: Springer.
Fink, G. A. (2008). *Markov models for pattern recognition.* Berlin, Heidelberg: Springer.
Flanagan, D. (1996). *Java in a nutshell: A desktop quick reference for Java programmers.* Sebastopol, CA, USA: O'Reilly & Associates, Inc.
Fleiss, J. L. (1971). Measuring nominal scale agreement among many raters. *Psychological Bulletin,76*, 378–382.
Freund, Y., & Schapire, R. E. (1995). A decision-theoretic generalization of on-line learning and an application to boosting. In *Proceedings of the Second European Conference on Computational Learning Theory* (pp. 23–37). London, UK: Springer.
Gamma, E., Helm, R., Johnson, R., & Vlissides, J. (1995). *Design patterns: Elements of reusable object-oriented software.* Boston, MA, USA: Addison-Wesley Longman Publishing Co., Inc.
Gödde, F., Möller, S., Engelbrecht, K.-P., Kühnel, C., Schleicher, R., Naumann, A. et al. (2008). Study of a Speech-based Smart Home System with Older Users. In *Proceedings of International Workshop on Intelligent User Interfaces for Ambient Assisted Living (IUI4AAL 2008), Canary Islands, Jan. 2008* (pp. 17–22). Stuttgart: Frauenhofer IRB Verlag.
Gorin, A. L., Parker, B. A., Sachs, R. M., & Wilpon, J. G. (1996). How may i help you? In *Interactive Voice Technology for Telecommunications Applications, 1996. Proceedings., Third IEEE Workshop on* (pp. 57–60).
Gorin, A. L., Riccardi, G., & Wright, J. H. (1997). How may i help you? *Journal of Speech Communication,23*(1–2), 113–127.
Goronzy, S. (2002). *Robust adaptation to non-native accents in automatic speech recognition. Lecture notes in computer science.* Berlin: Springer.
Gruber, O., Hargrave, B. J., McAffer, J., Rapicault, P., & Watson, T. (2005). The eclipse 3.0 platform: Adopting osgi technology. *IBM Systems Journal, 44*, 289–300.
Gwet, K. L. (2010). *Handbook of inter-rater reliability* (2nd ed.). Advanced Analytics, LLC.

Hajdinjak, M., & Mihelic, F. (2006). The paradise evaluation framework: Issues and findings. *Computational Linguistics,32*(2), 263–272.

Hajdinjak, M., & Mihelic, F. (2007). A wizard-of-oz system evaluation study. In *Proceedings of the 10th international conference on text, speech and dialogue, TSD'07* (pp. 532–539). Berlin, Heidelberg: Springer.

Hall, J. L. (1998). *The digital signal processing handbook*, chapter Auditory Psychophysics for Coding Applications. Number 39. New York: CRC Press, IEEE Press.

Hara, S., Kitaoka, N., & Takeda, K. (2010). Estimation method of user satisfaction using n-gram-based dialog history model for spoken dialog system. In N. C. C. Chair, K. Choukri, B. Maegaard, J. Mariani, J. Odijk, S. Piperidis, M. Rosner, & D. Tapias, (Eds.), *Proceedings of the Seventh conference on International Language Resources and Evaluation (LREC'10), Valletta, Malta*. European Language, Resources Association (ELRA).

Heinroth, T., Denich, D., & Schmitt, A. (2010a). Owlspeak—adaptive spoken dialogue within intelligent environments. In *IEEE PerCom Workshop Proceedings*.

Heinroth, T., Denich, D., Schmitt, A., & Minker, W. (2010b). Efficient spoken dialogue domain representation and interpretation. In *Proceedings of the Seventh Conference on International Language Resources and Evaluation (LREC'10), Valetta, Malta*. European Language, Resources Association (ELRA).

Heinroth, T., Schmitt, A., & Bertrand, G. (2009). Enhancing speech dialogue technologies for ambient intelligent environments. In *5th International Conference on Intelligent Environments (IE' 09)*.

Herm, O., Schmitt, A., & Liscombe, J. (2008). When calls go wrong: How to detect problematic calls based on log-files and emotions? In *Proceedings of the International Conference on Speech and Language Processing (ICSLP) Interspeech 2008* (pp. 463–466).

Higashinaka, R., Minami, Y., Dohsaka, K., & Meguro, T. (2010a). Issues in predicting user satisfaction transitions in dialogues: Individual differences, evaluation criteria, and prediction models. In G. Lee, J. Mariani, W. Minker & S. Nakamura (Eds.), *Spoken Dialogue Systems for Ambient Environments, volume 6392 of Lecture Notes in Computer Science* (pp. 48–60). Berlin, Heidelberg: Springer. doi:10.1007/978-3-642-16202-2_5.

Higashinaka, R., Minami, Y., Dohsaka, K., & Meguro, T. (2010b). Modeling user satisfaction transitions in dialogues from overall ratings. In *Proceedings of the SIGDIAL 2010 Conference* (pp. 18–27). Tokyo, Japan: Association for Computational Linguistics.

Hone, K. S., & Graham, R. (2000). Towards a tool for the subjective assessment of speech system interfaces (sassi). *Natural Language Engineering,6*(3–4), 287–303.

Huang, X., Acero, A., & Hon, H.-W. (2001). *Spoken language processing: A guide to theory, algorithm, and system development* (1st ed.). Upper Saddle River, NJ, USA: Prentice Hall PTR.

Hughes, G. (1968). On the mean accuracy of statistical pattern recognizers. *IEEE Transactions on Information Theory, 14*(1), 55–63.

ISO (1998). *Ergonomic requirements for office work with visual display terminals (VDTs), Part 11: Guidance on usability*. International Standardization Organization (ISO).

ITU (1994). *Terms and definitions related to quality of service and network performance including dependability*. ITU-T Recommendation E.800, International Telecommunication Union, Geneva, Switzerland.

ITU (2005). *Parameters describing the interaction with spoken dialogue systems*. ITU-T Recommendation Supplement 24 to P-Series, International Telecommunication Union, Geneva, Switzerland. Based on ITU-T Contr. COM 12–17 (2009).

ITU (2007). *Vocabulary for performance and quality of service*. ITU-T Amendment Amendment 1 to P.10/G.100, International Telecommunication Union, Geneva, Switzerland.

Ives, B., Olson, M. H., & Baroudi, J. J. (1983). The measurement of user information satisfaction. *Communication ACM,26*, 785–793.

James, W. (1884). What is an emotion? *Mind,9*(34), 188–205.

Joanes, D. N., & Gill, C. A. (1998). Comparing measures of sample skewness and kurtosis. *Journal of the Royal Statistical Society Series D (The Statistician),47*(1), 183–189.

Jurafsky, D., & Martin, J. H. (2008). *Speech and language processing* (2nd ed.). Upper Saddle River: Prentice Hall.

Kamel, M. S., & Karray, F. (2007). Speech emotion recognition using gaussian mixture vector. In *Proceedings of the International Conference on Acoustics, Speech, and Signal Processing (ICASSP)*.

Kamm, C. A., Litman, D., & Walker, M. (1998). From novice to expert: The effect of tutorials on user expertise with spoken dialogue systems. In *Proceedings of the International Conference on Spoken Language Processing (ICSLP)* (pp. 1211–1214).

Kim, H.-C., Pang, S., Je, H.-M., Kim, D., & Bang, S.-Y. (2002). Support vector machine ensemble with bagging. In S.-W. Lee & A. Verri (Eds.), *Pattern recognition with support vector machines, volume 2388 of Lecture Notes in Computer Science* (pp. 131–141). Berlin, Heidelberg: Springer. doi:10.1007/3-540-45665-1_31.

Kim, K., Bang, S., & Kim, S. (2004). Emotion recognition system using short-term monitoring of physiological signals. *Medical and Biological Engineering and Computing,42*(3), 419–427.

Kim, W. (2007). Online call quality monitoring for automating agent-based call centers. In *Proceedings of the International Conference on Speech and Language Processing (ICSLP)*.

Kinsler, L. E., Frey, A. R., Coppens, A. B., & Sanders, J. V. (1982). Fundamentals of acoustics (3rd ed.). New York: Wiley.

Kuncheva, L. I. (2004). *Combining pattern classifiers: Methods and algorithms*. New York: Wiley-Interscience.

Lamel, L., Minker, W., & Paroubek, P. (2000). Towards best practice in the development and evaluation of speech recognition components of a spoken language dialog system. *Natural Language Engineering,6*, 305–322.

Landis, J. R., & Koch, G. G. (1977). The measurement of observer agreement for categorical data. *Biometrics,33*(1), 159–174.

Langkilde, I., Walker, M., Wright, J., Gorin, A., & Litman, D. (1999). Automatic prediction of problematic human-computer dialogues in how may i help you. In *Proceedings of the IEEE Workshop on Automatic Speech Recognition and Understanding, ASRU99* (pp. 369–372).

Langley, P. (1997). Machine learning for adaptive user interfaces. In *Proceedings of the 21st Annual German Conference on Artificial Intelligence: Advances in Artificial Intelligence* (pp. 53–62). London, UK: Springer.

Langley, P., Thompson, C., Elio, R., & Haddadi, A. (1999). An adaptive conversational interface for destination advice. In *Proceedings of the Third International Workshop on Cooperative Information Agents* (pp. 347–364).

Larcker, D., & Lessig, V. P. (1980). Perceived usefulness of information: A psychometric examination. *Decision Sciences, 11*, 121–134.

Larson, J. A. (2002). *Voicexml: Introduction to developing speech applications: Version*. Upper Saddle River, NJ, USA: Prentice Hall PTR.

Larsson, S., & Traum, D. (2000). Information state and dialogue management in the trindi dialogue move engine. *Natural Language Engineering Special Issue,6*, 323–340.

Lee, C. M., & Narayanan, S. S. (2005). Toward detecting emotions in spoken dialogs. *IEEE Transactions on Speech and Audio Processing,13*(2), 293–303.

Lee, C. M., Narayanan, S. S., & Pieraccini, R. (2001). Recognition of negative emotions from the speech, signal. In *Proceedings of IEEE Workshop on Automatic Speech Recognition and Understanding* (pp. 240–243).

Lee, C. M., Narayanan, S. S., & Pieraccini, R. (2002). Combining acoustic and language information for emotion. In *Proceedings of International Conference on Speech and Language Processing (ICSLP) 2002*

Lee, S., Yildirim, S., Kazemzadeh, A., & Narayanan, S. S. (2005). An articulatory study of emotional speech production. In *Proceedings of European Conference on Speech and Language Processing (EUROSPEECH)*.

Lemon, O., & Pietquin, O. (2011). Introduction to special issue on machine learning for adaptivity in spoken dialogue systems. *ACM Transactions on Speech and Language Processing,7*(3), 1–15

Levin, E., & Pieraccini, R. (2006). Value-based optimal decision for dialog systems. In *Proceedings of Spoken Language Technology Workshop 2006* (pp. 198–201).

Lindgaard, G., & Dudek, C. (2003). What is this evasive beast we call user satisfaction? *Interacting with Computers,15*(3), 429–452.

Liscombe, J., Riccardi, G., & Hakkani-Tür, D. (2005). Using context to improve emotion detection in spoken dialog systems. In *Proceedings of International Conference on Speech and Language Processing (ICSLP)* (pp. 1845–1848).

Litman, D., Hirschberg, J. B., & Swerts, M. (2000). Predicting automatic speech recognition performance using prosodic cues. In *Proceedings of the 1st North American chapter of the Association for Computational Linguistics conference* (pp. 218–225). San Francisco, CA, USA: Morgan Kaufmann Publishers Inc.

Litman, D., & Pan, S. (1999). Empirically evaluating an adaptable spoken dialogue system. In *Proceedings of the 7th International Conference on User Modeling* (pp. 55–64).

Litman, D., & Pan, S. (2002). Designing and evaluating an adaptive spoken dialogue system. *User Modeling and User-Adapted Interaction,12*(2–3), 111–137.

Litman, D., Walker, M., & Kearns, M. S. (1999). Automatic detection of poor speech recognition at the dialogue level. In *Proceedings of the 37th annual meeting of the Association for Computational Linguistics on Computational Linguistics* (pp. 309–316) Morristown, NJ, USA: Association for Computational Linguistics.

López-Cózar, R., Callejas, Z., Kroul, M., Nouza, J., & Silovský, J. (2008). Two-level fusion to improve emotion classification in spoken dialogue systems. In *Text, Speech and Dialogue* (pp. 617–624). Berlin: Springer.

López-Cózar, R., Callejas, Z., & McTear, M. (2006). Testing the performance of spoken dialogue systems by means of an artificially simulated user. *Artificial Intelligence Review,26*, 291–323.

López-Cózar, R., De la Torre, A., Segura, J. C., & Rubio, A. J. (2003). Assessment of dialogue systems by means of a new simulation technique. *Speech Communication,40*, 387–407.

López-Cózar, R., de la Torre, Á., Segura, J. C., Rubio, A. J., & Sánchez, V. E. (2002). Testing dialogue systems by means of automatic generation of conversations. *Interacting with Computers,14*(5), 521–546.

López-Cózar, R., Rubio, A. J., Verdejo, J. E. D., & la Torre, A. D. (2000). Evaluation of a dialogue system based on a generic model that combines robust speech understanding and mixed-initiative control. In *Proceedings of Language Resources and Evaluation (LREC)*.

Manning, C. D., Raghavan, P., & Schtze, H. (2008). *Introduction to information retrieval*. New York, NY, USA: Cambridge University Press.

Martin, J. C., Caridakis, G., Devillers, L., Karpouzis, K., & Abrilian, S. (2006). Manual annotation and automatic image processing of multimodal emotional behaviors: Validating the annotation of tv interviews. *Personal Ubiquitous Computing,13*(1), 69–76.

McAffer, J., & Lemieux, J.-M. (2005). *Eclipse rich client platform: Designing, coding, and packaging Java(TM) applications*. Reading: Addison-Wesley Professional.

McAffer, J., Lemieux, J.-M., & Aniszczyk, C. (2010). *Eclipse rich client platform* (2nd ed.). Reading: Addison-Wesley Professional.

McKelvie, D., Isard, A., Mengel, A., Møller, M. B., Grosse, M., & Klein, M. (2001). The mate workbench—an annotation tool for xml coded speech corpora. *Speech Communication, 33*(1–2):97–112. Speech Annotation and Corpus Tools.

McTear, M. F. (2004). *Spoken dialogue technology: Towards the conversational user interface*. London: Springer.

Metze, F., Ajmera, J., Englert, R., Bub, U., Burkhardt, F., Stegmann, J. et al. (2007). Comparison of four approaches to age and gender recognition. In *Proceedings of the International Conference on Acoustics, Speech, and Signal Processing (ICASSP)* (Vol. 1).

Metze, F., Englert, R., Bub, U., Burkhardt, F., & Stegmann, J. (2008). *Getting closer: Tailored human-computer speech dialog*. Universal Access in the Information Society.

Metze, F., Polzehl, T., & Wagner, M. (2009). Fusion of acoustic and linguistic speech features for emotion detection. In *Proceedings of International Conference on Semantic Computing (ICSC 2009)* (Vol. 1). IEEE, Berleley, CA, USA

Mierswa, I., Wurst, M., Klinkenberg, R., Scholz, M., & Euler, T. (2006). Yale: Rapid prototyping for complex data mining tasks. In L. Ungar, M. Craven, D. Gunopulos & T. Eliassi-Rad (Eds.), *KDD '06: Proceedings of the 12th ACM SIGKDD International Conference on Knowledge Discovery and Data Mining* (pp. 935–940). New York, NY, USA: ACM.

Miller, S., Bobrow, R., Ingria, R., & Schwartz, R. (1994). Hidden understanding models of natural language. In *Proceedings of the 32nd annual meeting on Association for Computational Linguistics* (pp. 25–32). Morristown, NJ, USA: Association for Computational Linguistics.

Minker, W., Waibel, A., & Mariani, J. (1999). *Stochastically-based semantic analysis*. Norwell, MA, USA: Kluwer Academic Publishers.

Müller, C., & Burkhardt, F. (2007). Combining short-term cepstral and long-term prosodic features for automatic recognition of speaker age. In *Proceedings of International Conference on Speech and Language Processing (ICSLP)*. New York: ISCA.

Molinaro, A. (2005). *SQL cookbook (cookbooks (O'Reilly))*. Sebastopol: O'Reilly Media, Inc.

Möller, S. (2005a). Neue itu-t-empfehlungen zur evaluierung telefonbasierter sprachdialogdienste. In A. B. Cremers, R. Manthey, P. Martini & V. Steinhage (Eds.), *INFORMATIK 2005—Informatik LIVE! Band 2, Beiträge der 35. Jahrestagung der Gesellschaft für Informatik e.V. (GI), Bonn, volume 68 of LNI*. Schweiz: Gesellschaft für Informatik.

Möller, S. (2005b). *Quality of telephone-based spoken dialogue systems*. New York: Springer.

Möller, S., Engelbrecht, K.-P., Kühnel, C., Wechsung, I., & Weiss, B. (2009). A taxonomy of quality of service and quality of experience of multimodal human-machine interaction. In *Proceedings of the International Workshop on Quality of Multimedia Experience, 2009. QoMEx 2009*. (pp. 7–12).

Möller, S., Engelbrecht, K.-P., & Schleicher, R. (2008). Predicting the quality and usability of spoken dialogue services. *Speech Communication,50*(8–9), 730–744.

Morrison, D., Wang, R., & De Silva, L. C. (2007). Ensemble methods for spoken emotion recognition in call-centres. *Speech Communication,49*, 98–112.

Mowafey, S., Schmitt, A., Hagras, H., & Minker, W. (2009). Creating an ambient intelligent environment with an emotion-aware system. In V. Callaghan, A. Kameas, A. Reyes, D. Royo & M. Weber (Eds.), *Proceedings of the 5th International Conference on Intelligent Environments* (pp. 236–246). Amsterdam: IOS Press.

Müller, C., & Wasinger, R. (2002). Adapting multimodal dialog for the elderly. *ABIS Workshop on Personalization for the Mobile World* (pp. 31–34). Germany: Hannover.

Müller, C., Wittig, F., & Baus, J. (2003). Exploiting speech for recognizing elderly users to respond to their special needs. In *Proceedings of the Eighth European Conference on Speech Communication and Technology (Eurospeech 2003)* (pp. 1305–1308).

Nisimura, R., Omae, S., Kawahara, H., & Irino, T. (2006). Analyzing dialogue data for real-world emotional speech classification. In *Proceedings of International Conference on Speech and Language Processing (ICSLP)*.

Nobuo, S., & Yasunari, O. (2007). Emotion recognition using mel-frequency cepstral coefficients. *Information and Media Technologies,2*, 835–848.

Oshry, M., Auburn, R., Baggia, P., Bodell, M., Burke, D., Burnett, D. et al. (2007). *Voice extensible markup language (voicexml) version 2.1*. Technical report, W3C—Voice Browser Working Group.

Paek, T., & Horvitz, E. (2004). Optimizing automated call routing by integrating spoken dialog models with queuing models. In *HLT-NAACL* (pp. 41–48).

Pantic, M., & Rothkrantz, L. J. M. (2003). Toward an affect-sensitive multimodal human-computer interaction. In *Proceedings of the IEEE* (pp. 1370–1390).

Paris, C. L. (1988). *The use of explicit user models in text generation: Tailoring to a user's level of expertise*. PhD thesis, New York, NY, USA. Order No: GAX88-15690.

Petrushin, V. A. (1999). Emotion in speech: Recognition and application to call centers. In *Artificial Neural Networks in Engineering (ANNIE '99), St. Louis* (pp. 7–10).

Pfister, B., & Kaufmann, T. (2008). *Sprachverarbeitung: Grundlagen und Methoden der Sprachsynthese und Spracherkennung*. Reihe: Springer-Lehrbuch, Springer.

Picard, R. (1997). *Affective computing*. Cambridge: MIT Press.

Picard, R. W. (2000). Toward computers that recognize and respond to user emotion. *IBM System Journal,39*, 705–719.

Pieraccini, R., & Huerta, J. (2005). Where do we go from here? research and commercial spoken dialog systems. In *Proceedings of the 6th SIGdial Workshop on Discourse and Dialog* (pp. 1–10).

Pieraccini, R., & Lubensky, D. (2005). Spoken language communication with machines: The long and winding road from research to business. In M. Ali & F. Esposito (Eds.), *Innovations in Applied Artificial Intelligence, volume 3533 of Lecture Notes in Computer Science* (pp. 6–15). Berlin, Heidelberg: Springer. doi:10.1007/11504894_2.

Pieraccini, R., Suendermann, D., Dayanidhi, K., & Liscombe, J. (2009). Are we there yet? research in commercial spoken dialog systems. In V. Matoušek & P. Mautner (Eds.), *Text, Speech and Dialogue, volume 5729 of Lecture Notes in Computer Science* (pp. 3–13). Berlin, Heidelberg: Springer. doi:10.1007/978-3-642-04208-9_3.

Pittermann, J., Pittermann, A., & Minker, W. (2009). *Handling emotions in human-computer dialogues*. Dordrecht, The Netherlands: Springer.

Pittermann, J., & Schmitt, A. (2008). Integrating linguistic cues into speech-based emotion recognition. In *4th IET International Conference on Intelligent, Environments, Seattle (USA)*.

Pittermann, J., Schmitt, A., & Minker, W. (2008). Comparing evaluation criteria for (automatic) emotion recognition. In *4th IET International Conference on Intelligent, Environments, Seattle (USA)* (pp. 1–4).

Platt, J. C. (1999). *Fast training of support vector machines using sequential minimal optimization* (pp. 185–208). Cambridge, MA, USA: MIT Press.

Plutchik, R. (1980). *Emotion: A psychoevolutionary synthesis*. New York: Harper & Row.

Polzehl, T., Metze, F., & Schmitt, A. (2010a). Factors for linguistic and prosodic emotion recognition. In *Fortschritte der Akustik—DAGA 2010*. DEGA e.V.

Polzehl, T., Schmitt, A., & Metze, F. (2009a). Comparing features for acoustic anger classification in german and english ivr portals. In *First International Workshop on Spoken Dialogue Systems (IWSDS)*.

Polzehl, T., Schmitt, A., & Metze, F. (2010b). Approaching multilingual emotion recognition from speech—on language dependency of acoustic/prosodic features for anger detection. In *Proceedings of the Fifth International Conference on Speech Prosody, 2010. Speech Prosody 2010*.

Polzehl, T., Schmitt, A., & Metze, F. (2011). Salient features for anger recognition in german and english ivr portals. In W. Minker, G. G. Lee, S. Nakamura & J. Mariani (Eds.), *Spoken dialogue systems technology and design* (pp. 83–105). New York: Springer. doi: 10.1007/978-1-4419-7934-6S_4.

Polzehl, T., Schmitt, A., Metze, F., & Wagner, M. (2010c). Anger recognition in speech using acoustic and linguistic cues. *Speech Communication*, Special Issue: Sensing Emotion and Affect—Facing Realism in Speech Processing.

Polzehl, T., Sundaram, S., Ketabdar, H., Wagner, M., & Metze, F. (2009b). Emotion classification in children's speech using fusion of acoustic and linguistic features. In *Proceedings of the International Conference on Speech and Language Processing (ICSLP) Interspeech 2009*.

Press, W.H., Teukolsky, S., Vetterling, W., & Flannery, B. (1992). *Numerical recipes in C* (2nd ed.). New York: Cambridge University Press.

Rabiner, L. R. (1990). *A tutorial on hidden Markov models and selected applications in speech recognition*. San Francisco, CA, USA: Morgan Kaufmann Publishers Inc.

Rabiner, L. R., & Juang, B.-H. (1993). *Fundamentals of speech recognition*. Upper Saddle River: Prentice Hall PTR.

Rabiner, L. R., & Sambur, M. R. (1975). An algorithm for determining the endpoints of isolated utterances. *The Bell System Technical Journal,56*, 297–315.

Raux, A., Bohus, D., Langner, B., Black, A. W., & Eskenazi, M. (2006). Doing research on a deployed spoken dialogue system: One year of let's go! experience. In *Proceedings of the International Conference on Speech and Language Processing (ICSLP).*

Raux, A., & Eskenazi, M. (2004). Non-native users in the let's go! spoken dialogue system: Dealing with linguistic mismatch. In *HLT-NAACL* (pp. 217–224).

Raux, A., Langner, B., Bohus, D., Black, A. W., & Eskenazi, M. (2005). Let's go public! taking a spoken dialog system to the real world. In *Proceedings of International Conference on Speech and Language Processing (ICSLP).*

Reeves, B., & Nass, C. (1996). *The media equation: How people treat computers, television, and new media like real people and places*. New York, NY, USA: Cambridge University Press.

Reiter, E., & Dale, R. (2000). *Building natural language generation systems*. New York, NY, USA: Cambridge University Press.

Riccardi, G., & Gorin, A. (2000). Stochastic language adaptation over time and state in natural spoken dialog systems. *IEEE Transactions on Speech and Audio Processing, 8*(1), 3–10

Riccardi, G., Gorin, A., Ljolje, A., & Riley, M. (1997). A spoken language system for automated call routing. In *Proceedings of the 1997 IEEE International Conference on Acoustics, Speech, and Signal Processing (ICASSP), ICASSP '97* (pp. 1143–1147). Washington, DC, USA: IEEE Computer Society.

Rieser, V, & Lemon, O. (2011). *Reinforcement Learning for Adaptive Dialogue Systems: A Data-driven Methodology for Dialogue Management and Natural Language Generation*. Theory and Applications of Natural Language Processing, Springer.

Russell, J. A. (1980). A circumplex model of affect. *Journal of Personality and Social Psychology,39*(6), 1161–1178.

Schapire, R. E., & Singer, Y. (2000). Boostexter: A boosting-based system for text categorization. *Machine Learning,39*, 135–168. doi:10.1023/A:1007649029923.

Scherer, K. R. (2000). Psychological models of emotion. In J. C. Borod (Ed.), *The neuropsychology of emotion* (pp. 137–166). Oxford, New York, Estados Unidos: Oxford University Press US.

Scherer, K. R. (2005). What are emotions? and how can they be measured? *Social Science Information,44*(4), 695–729.

Schlosberg, H. (1954). Three dimensions of emotion. *Psychological Review,61*(2), 81–88.

Schmitt, A., Bertrand, G., Heinroth, T., & Liscombe, J. (2010a). Witchcraft: A workbench for intelligent exploration of human computer conversations. In *Proceedings of the International Conference on Language Resources and Evaluation (LREC), Valetta, Malta.*

Schmitt, A., Hank, C., & Liscombe, J. (2008a). Detecting problematic calls with automated agents. In *4th IEEE Tutorial and Research Workshop Perception and Interactive Technologies for Speech-Based Systems* (pp. 72–80). Irsee, Germany: Springer.

Schmitt, A., Heinroth, T., & Bertrand, G. (2009a). Towards emotion, age- and gender-aware voicexml applications. In *Proceedings of the 5th International Conference on Intelligent, Environments (IE'09).*

Schmitt, A., Heinroth, T., & Liscombe, J. (2009). On nomatchs, noinputs and bargeins: Do non-acoustic features support anger detection? In *Proceedings of the 10th Annual SIGDIAL Meeting on Discourse and Dialogue, SigDial Conference 2009*. London, UK: Association for Computational Linguistics.

Schmitt, A., Minker, W., & Sharaf, N. (2010b). Advances in the witchcraft workbench project. In *Proceedings of the SIGDIAL 2010 Conference* (pp. 261–264). Tokyo, Japan: Association for Computational Linguistics.

Schmitt, A., Pieraccini, R., & Polzehl, T. (2010c). *Advances in Speech Recognition: Mobile Environments, Call Centers and Clinics, chapter 'For Heaven's sake, gimme a live person!'*. Designing Emotion-Detection Customer Care Voice Applications in Automated Call Cent, Springer.

Schmitt, A., & Polzehl, T. (2010). Modeling a-priori likelihoods for angry user turns with hidden markov models. In *Proceedings of the Fifth International Conference on Speech Prosody, 2010*. Speech Prosody 2010.

Schmitt, A., Polzehl, T., & Liscombe, J. (2010d). The influence of the utterance length on the recognition of aged voices. In *Proceedings of the International Conference on Language Resources and Evaluation (LREC), Valetta, Malta.*

Schmitt, A., Polzehl, T., & Minker, W. (2010e). Facing reality: Simulating deployment of anger recognition in ivr systems. In G. G. Lee, J. Mariani, W. Minker & S. Nakamura (Eds.), *Proceedings of the Second International Workshop on Spoken Dialogue Systems (IWSDS)* (pp. 122–131).

Schmitt, A., Schatz, B., & Minker, W. (2011a). Modeling and predicting quality in spoken human-computer interaction. In *Proceedings of the SIGDIAL 2011 Conference, Oregon, Portland. USA*. Association for Computational Linguistics.

Schmitt, A., Schatz, B., & Minker, W. (2011b). A statistical approach for estimating user satisfaction in spoken human-machine interaction. In *Proceedings of the IEEE Jordan Conference on Applied Electrical Engineering and Computing Technologies (AEECT)*, Amman, Jordan, IEEE.

Schmitt, A., Scholz, M., Minker, W., Liscombe, J., & Sündermann, D. (2010f). Is it possible to predict task completion in automated troubleshooters? In T. Kobayashi, K. Hirose & S. Nakamura (Eds.), *Proceedings of the International Conference on Speech and Language Processing (ICSLP)* (pp. 94–97).

Schmitt, A., Tschaffon, U., Heinroth, T., & Minker, W. (2010g). Inter-labeler agreement for anger detection in interactive voice response systems. In *Proceedings of the 6th International Conference on Intelligent, Environments (IE'10)* (pp. 112–115).

Schmitt, A., Zaykovskiy, D., & Minker, W. (2008b). Speech recognition for mobile devices. *International Journal of Speech Technology,11*, 63–72. doi:10.1007/s10772-009-9036-6.

Schmitt, A., Zgorzelski, A., & Minker, W. (2011c). Tackling a shilly-shally classifier for predicting task success in spoken dialogue interaction. In *Proceedings of the International Conference on Speech and Language Processing (ICSLP)* (pp. 805–808).

Scholkopf, B., & Smola, A. J. (2001). *Learning with kernels: Support vector machines, regularization, optimization, and beyond*. Cambridge, MA, USA: MIT Press.

Schröder, M., & Trouvain, J. (2003). The german text-to-speech synthesis system mary: A tool for research, development and teaching. *International Journal of Speech Technology,6*, 365–377.

Schuller, B. (2006). *Automatische Emotionserkennung aus sprachlicher und manueller Interaktion*. Dissertation: Technische Universität München, München.

Schuller, B., Seppi, D., Batliner, A., Maier, A., & Steidl, S. (2007). Towards more reality in the recognition of emotional speech. *IEEE International Conference on Acoustics Speech and Signal Processing (ICASSP), 4*(101), 3–6.

Schuller, B., Steidl, S., & Batliner, A. (2009). The interspeech 2009 emotion challenge. In *Proceedings of the International Conference on Speech and Language Processing (ICSLP)*.

Schuller, B., Steidl, S., Batliner, A., Burkhardt, F., Devillers, L., Müller, C. et al. (2010). The interspeech 2010 paralinguistic challenge. In *Proceedings of the International Conference on Speech and Language Processing (ICSLP)* (pp. 2794–2797).

Schuller, B., Steidl, S., Batliner, A., Schiel, F., & Krajewski, J. (2011). The interspeech 2011 speaker state challenge. In *Proceedings of the International Conference on Speech and Language Processing (ICSLP)*.

Schürmann, J. (1996). *Pattern classification—a unified view of statistical and neural approaches*. New Year: Wiley.

Shafran, I. (2005). A comparison of classifiers for detecting emotion from speech. In *IEEE International Conference on Acoustics, Speech and Signal Processing*.
Shafran, I., Riley, M., & Mohri, M. (2003). Voice signatures. In *Automatic Speech Recognition and Understanding, 2003. ASRU '03. 2003 IEEE Workshop on* (pp. 31–36).
Smith, B. (2007). *A quick guide to gplv3*. Technical report, Free Software Foundation, Inc.
Spearman, C. (1904). The proof and measurement of association between two things. *American Journal of Psychology,15*, 88–103.
Steidl, S. (2009). *Automatic classification of emotion-related user states in spontaneous children's speech* (1st ed.). Berlin: Logos Verlag.
Steidl, S., Levit, M., Batliner, A., Nöth, E., & Niemann, H. (2005). "Of all things the measure is man"—classification of emotions and inter-labeler consistency. In IEEE (Ed.), *Proceedings of the International Conference on Acoustics, Speech, and Signal Processing (ICASSP)* (Vol. 1, pp. 317–320), 3833 S. Texas Ave., Ste. 221 Bryan, TX 77802–4015.
Stibbard, R. (2001). *Vocal Expression of Emotions in Non-Laboratory Speech: An Investigation of the Reading/Leeds Emotion in Speech Project Annotation Data*. PhD thesis, University of Reading, United Kingdom.
Stolcke, A., Ries, K., Coccaro, N., Shriberg, E., Bates, R., Jurafsky, D., et al. (2000). Dialogue act modeling for automatic tagging and recognition of conversational speech. *Computational Linguistics,26*, 339–373.
Strauss, P.-M., & Minker, W. (2010). *Proactive spoken dialogue interaction in multi-party environments*. Boston, USA: Springer.
Ultes, S., Heinroth, T., Schmitt, A., & Minker, W. (2011a). A theoretical framework for a user-centered spoken dialog manager. In *Proceedings of the Paralinguistic Information and its Integration in Spoken Dialogue Systems Workshop* (pp. 241–246). Berlin: Springer.
Ultes, S., Schmitt, A., & Minker, W. (2011b). Attention, sobriety checkpoint! can humans determine by means of voice, if someone is drunk... and can automatic classifiers compete? In *Proceedings of the 12th Annual Conference of the International Speech Communication Association (INTERSPEECH 2011)*.
van den Bosch, A., Krahmer, E., & Swerts, M. (2001). Detecting problematic turns in human-machine interactions: Rule-induction versus memory-based learning approaches. In *ACL '01: Proceedings of the 39th Annual Meeting on Association for Computational Linguistics* (pp. 82–89). Morristown, NJ, USA: Association for Computational Linguistics.
Viterbi, A. (1967). Error bounds for convolutional codes and an asymptotically optimum decoding algorithm. *IEEE Transactions on Information Theory,13*(2), 260–269.
Vlasenko, B., Schuller, B., Mengistu, K. T., Rigoll, G., & Wendemuth, A. (2008). Balancing spoken content adaptation and unit length in the recognition of emotion and interest. In *Proceedings of International Conference on Speech and Language Processing (ICSLP)* (pp. 805–808).
Vlasenko, B., & Wendemuth, A. (2009). Processing affected speech within human machine interaction. In *Proceedings of International Conference on Speech and Language Processing (ICSLP)*
Walker, M., Kamm, C., & Litman, D. (2000). Towards developing general models of usability with paradise. *Natural Language Engineering,6*(3–4), 363–377.
Walker, M., Langkilde-Geary, I., Hastie, H. W., Wright, J., & Gorin, A. (2002). Automatically training a problematic dialogue predictor for a spoken dialogue system. *Journal of Artificial Intelligence Research,16*, 293–319.
Walker, M., Litman, D., Kamm, A. A., & Abella, A. (1998). Evaluating spoken dialogue agents with PARADISE: Two case studies. *Computer Speech and Language, 12*(4): 317–347.
Walker, M., Litman, D., Kamm, C. A., & Abella, A. (1997). Paradise: A framework for evaluating spoken dialogue agents. In *Proceedings of the eighth conference on European chapter of the Association for Computational Linguistics* (pp. 271–280). Morristown, NJ, USA: Association for Computational Linguistics.
Webb, A. R. (2002). *Statistical pattern recognition* (2nd ed.). New York: Wiley.

Weiss, B., Wechsung, I., Naumann, A., & Möller, S. (2008). Subjective evaluation method for speech-based uni- and multimodal applications. In *Proceedings of the 4th IEEE tutorial and research workshop on Perception and Interactive Technologies for Speech-Based Systems: Perception in Multimodal Dialogue Systems, PIT '08* (pp. 285–288). Berlin, Heidelberg: Springer.

Williams, J. D., & Young, S. J. (2007). Partially observable markov decision processes for spoken dialog systems. *Computer Speech and Language,21*, 393–422.

Witten, I. H., & Frank, E. (2005). *Data mining: Practical machine learning tools and techniques* (2nd ed.). San Francisco: Morgan Kaufmann.

Yacoub, S., Simske, S., Lin, X., & Burns, J. (2003). Recognition of emotions in interactive voice response systems. In *Proceedings of European Conference on Speech and Language Processing (EUROSPEECH)* (pp. 1–4).

Yang, F., & Heeman, P. A. (2005). Dialogue view: an annotation tool for dialogue. In *Proceedings of HLT/EMNLP on Interactive Demonstrations* (pp. 20–21). Morristown, NJ, USA: Association for Computational Linguistics.

Young, S. J. (1994). The htk hidden markov model toolkit: Design and philosophy. *Entropic Cambridge Research Laboratory, Ltd.,2*, 2–44.

Young, S. J., Williams, J. D., Schatzmann, J., Stuttle, M. N., & Weilhammer, K. (2006). *D4.3: Bayes net prototype—the hidden information state dialogue manager*. Technical report, TALK—Talk and Look: Tools for Ambient Linguistic Knowledge, IST-507802, 6th FP.

Zablotskaya, K., Abbas, M., Zablotskiy, S., Walter, S., & Minker, W. (2011a). Measuring verbal intelligence using linguistic analysis. In *Intelligent Environments'11* (pp. 88–91).

Zablotskaya, K., Rahim, U., Zablotskiy, S., Walter, S., & Minker, W. (2011b). Conversation peculiarities of people with different verbal intelligence. In R. L.-C. Delgado & T. Kobayashi (Eds.), *Proceedings of the Paralinguistic Information and its Integration in Spoken Dialogue Systems Workshop* (pp. 157–163). New York: Springer.

Zaykovskiy, D., & Schmitt, A. (2007). Java to micro edition front-end for distributed speech recognition systems. In *The 2007 IEEE International Symposium on Ubiquitous Computing and Intelligence (UCI'07), Niagara Falls (Canada)*.

Zaykovskiy, D., & Schmitt, A. (2008a). Deploying dsr technology on today's mobile phones: A feasibility study. In *4th IEEE Tutorial and Research Workshop Perception and Interactive Technologies for Speech-Based Systems, Irsee (Germany)*.

Zaykovskiy, D., & Schmitt, A. (2008b). Java vs. symbian: A comparison of software-based DSR implementations on mobile phones. In *4th IET International Conference on Intelligent, Environments, Seattle (USA)*.

Zaykovskiy, D., & Schmitt, A. (2008c). Java vs. symbian: A comparison of software-based dsr implementations on mobile phones. In *4th IET International Conference on Intelligent, Environments, Seattle (USA)*.

Zaykovskiy, D., Schmitt, A., & Lutz, M. (2007). New use of mobile phones: Towards multimodal information access systems. In *3rd IET International Conference on Intelligent, Environments, Ulm (Germany)*.

Zgorzelski, A., Schmitt, A., Heinroth, T., & Minker, W. (2010). Repair strategies on trial: Which error recovery do users like best?. In *Proceedings of the International Conference on Speech and Language Processing (ICSLP)*.

Zweig, G., Siohan, O., Saon, G., Ramabhadran, B., Povey, D., Mangu, L. et al. (2006). Automated quality monitoring in the call center with asr and maximum entropy. In *Proceedings of IEEE International Conference on Acoustics, Speech and Signal Processing (ICASSP), 2006* (Vol. 1, pp. I–I).

Index

A. Schmitt and W. Minker, *Towards Adaptive Spoken Dialog Systems*,
DOI: 10.1007/978-1-4614-4593-7,

Printed by Publishers' Graphics LLC